Proof in Geometry

by A. I. Fetisov

WITH

Mistakes in Geometric Proofs

by Ya. S. Dubnov

Dover Publications, Inc.
Mineola, New York

Bibliographical Note

This Dover edition, first published in 2006, is an unabridged republication, in one volume, of *Proof in Geometry* by A. I. Fetisov and *Mistakes in Geometric Proofs* by Ya. S. Dubnov, both of which were originally published in English by D. C. Heath and Company, Boston, in 1963. *Proof in Geometry* was translated and adapted from the first Russian edition (1954) by Theodore M. Switz and Luise Lange. *Mistakes in Geometric Proofs* was translated and adapted from the second Russian edition (1955) by Alfred K. Henn and Olga A. Titelbaum.

Library of Congress Cataloging-in-Publication Data

Fetisov, A. I.
[O dokazatel'stve v geometrii. English]
Proof in geometry / by A. I. Fetisov. With Mistakes in geometric proofs / by Ya. S. Dubnov.
p. cm.
"This Dover edition . . . is an unabridged republication, in one volume, of Proof in geometry by A. I. Fetisov and Mistakes in geometric proofs by Ya. S. Dubnov, both of which were originally published in English by D.C. Heath and Company, Boston, in 1963"—T.p. verso.
ISBN-13: 978-0-486-45354-5
ISBN-10: 0-486-45354-5 (pbk.)
1. Axioms. 2. Logic, Symbolic and mathematical. 3. Fallacies (logic) 4. Geometry. I. Dubnov, IA. S. (IAkov Semenovich), 1887–1957. Oshibki v geometricheskikh dokazatel'stvakh. English. II. Title. III. Title: Mistakes in geometric proofs.

QA481.F433 2006
516—dc22

2006050208

Manufactured in the United States by Courier Corporation
45354502
www.doverpublications.com

Proof in Geometry

by A. I. Fetisov

PREFACE TO THE AMERICAN EDITION

THIS BOOKLET discusses the construction of geometric proofs and gives some criteria useful for determining whether or not a proof is logically correct and whether or not it actually proves what it was meant to prove. After some preliminary remarks on the role of axioms in geometry, there is a discussion of some common logical pitfalls responsible for invalid proofs—circular reasoning, assuming "obvious facts," examining only special cases, and so on. The discussion centers around sample invalid proofs that contain these logical errors. In each case the invalid proof is accompanied by a valid one, along with suggestions for avoiding the pitfall.

The last chapter discusses some of the axioms from Hilbert's famous set of axioms for Euclidean geometry. The properties of independence, completeness, and consistency are discussed for axiomatic systems in general.

This booklet can be read by anyone familiar with high school geometry.

CONTENTS

Introduction

1. FIRST STUDENT'S QUESTION

One day, at the beginning of the school year, I happened to overhear a conversation between two young girls. The older one had just begun the study of plane geometry. They were discussing their impressions of their lessons, teachers, girl friends, and the new subjects they were studying. The older girl was quite surprised at the lessons in geometry. "You would not believe it," she said. "The teacher came in, drew two congruent triangles on the board, and then spent the whole hour proving that they were—congruent! I don't see it. What's the use of doing that?" "But how are you going to recite in class?" asked the younger one. "I'll study it from the book . . . only it's hard to remember where to put all those letters . . . ."

That evening I overheard her muttering to herself repeatedly as she sat at the window diligently studying her geometry, "To prove it, place triangle $A'B'C'$ on triangle ABC . . . ." Unfortunately, I never found out how well she succeeded eventually in learning her geometry, but I think it may well have been difficult for her.

2. SECOND STUDENT'S QUESTION

A few days later, Tolya, my young neighbor across the hall, came to see me. He, too, has complaints about geometry. His homework assignment, after explanations given in class, was to study the theorem that in a triangle an exterior angle is larger than a nonadjacent interior angle. Showing me the figure from Kiselev's textbook[1] (Fig. 1), Tolya asked, "Why is it necessary to give a long and complicated proof, when the figure shows clearly that the exterior angle is obtuse, and the nonadjacent interior angles are

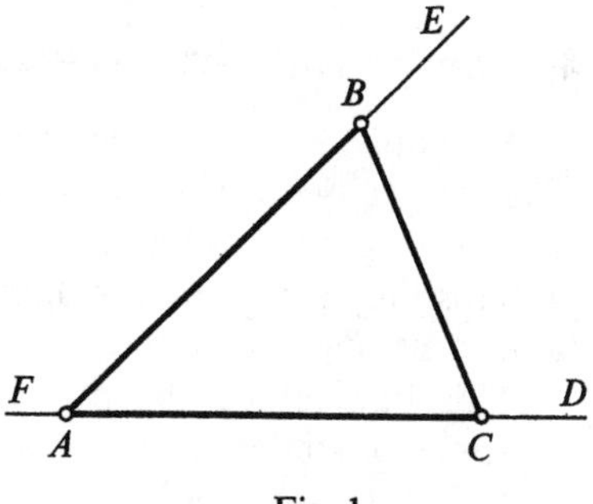

Fig. 1

[1] *Editor's note.* Under the centralized Russian school system the same "standard textbooks" are used in all schools. The standard texts in geometry to which repeated references are made in this book are by A. P. Kiselev and by H. A. Glagolev.

acute? An obtuse angle is always larger than an acute one," he argued. "That's clear without proof." So I explained to him why this proposition is not at all evident, and that there is good reason indeed to require proof for it.

3. THIRD STUDENT'S QUESTION

Again, a boy studying more advanced geometry recently showed me a paper of his on which, in his words, his grade had been "unjustly" lowered. The problem had been to determine the altitude of an isosceles trapezoid with bases 9 and 25 cm. long and one side 17 cm. long. To solve this problem he had inscribed a circle in the trapezoid stating that this was possible by virtue of the theorem that in any quadrilateral circumscribed about a circle the sums of the opposite sides are equal, which was true in the given trapezoid ($9 + 25 = 17 + 17$). He had then determined the altitude as the diameter of the circle inscribed in the isosceles trapezoid, which—as had been proved in a problem solved earlier—is the mean proportional between the two bases.

The solution seemed very simple and conclusive to him. But the teacher had rejected his reference to the theorem of the sums of the sides in a circumscribed quadrilateral as incorrect. This the boy could not see. He kept insisting, "But isn't it true that in a quadrilateral circumscribed about a circle the sums of the opposite sides are equal? Well, in this trapezoid the sum of the two bases is equal to that of the sides, which means that one can inscribe a circle in it. What is wrong with that?"

4. HOW TO FIND THE ANSWERS

We could give many similar examples showing that students often fail to see the need for a proof of what seems obvious to them, or regard proof as unduly complicated and cumbersome; or perhaps they accept as conclusive a proof which, on closer inspection, turns out to be false.

This booklet has been written to help answer the following questions of students:

1. What is a proof?
2. Why are proofs necessary?
3. How should a proof be constructed?
4. What propositions in geometry are accepted without proof?

1. What Is a Proof?

5. INDUCTION AND DEDUCTION

Let us ask ourselves, "What is a proof?" Suppose you are trying to convince a friend that the earth is spherical in shape. You tell him about the widening of the horizon as the observer rises above the surface of the earth, about voyages around the world, about the round shadow which the earth casts on the moon during a lunar eclipse, etc.

These statements, by which you seek to convince your friend, are called *arguments.* On what is the strength or the conclusiveness of an argument based? Let us look, for example, at the last of the above arguments. We claim that the earth must be round because its shadow is round. This assertion is based on the fact, which we know from experience, that all bodies that have a spherical form cast a round shadow, and conversely, that bodies have spherical form if they cast round shadows regardless of their position. Thus, in this case, we rely first of all on *facts*, on our own immediate experience regarding the properties of bodies in our everyday surroundings. Then we have recourse to a *deduction*, which in the case given is established in approximately the following manner:

"All bodies which in all different positions cast a round shadow have the shape of a sphere." "The earth, which during lunar eclipses occupies different positions in relation to the moon, always casts a round shadow on it." Conclusion: "Therefore, the earth has the shape of a sphere."

Let us take an example from physics. In the sixties of the last century, the English physicist Maxwell found that electromagnetic waves spread through space with the same velocity as light. This discovery led him to hypothesize that light is also an electromagnetic wave. But to prove the correctness of this hypothesis, it was necessary to show that the similarity between light waves and electromagnetic waves is not limited to their equal velocities of propagation. Rather it was necessary to adduce other weighty arguments to show the identical nature of both phenomena. Such arguments were forthcoming as a result of various experiments which showed the unquestionable influence of magnetic and electric fields on light emitted by various sources. A whole series of other facts was discovered which gave further evidence that light waves and electromagnetic waves are identical in nature.

Now let us turn to an example from arithmetic. We take any arbitrary odd numbers, square them, and subtract the number one from each square. For example,

$$7^2 - 1 = 48, \quad 11^2 - 1 = 120, \quad 5^2 - 1 = 24,$$
$$9^2 - 1 = 80, \quad 15^2 - 1 = 224,$$

and so forth. When we look at the resulting numbers, we notice that they have a property in common—each of them is exactly divisible by 8. Carrying out a few more such trials with other odd numbers and always finding the result to be divisible by 8, we might state tentatively, "The square of any odd number, diminished by one, gives a number which is a multiple of 8."

Since we are now speaking of *any* odd number, for a proof we must find arguments which serve for *any arbitrary* odd number. To do this, we first recall that any odd number can be expressed in the form $2n - 1$, where n is an integer. The square of an odd number, diminished by one, is then given by the expression $(2n - 1)^2 - 1$. Removing the parentheses, we get

$$(2n - 1)^2 - 1 = 4n^2 - 4n + 1 - 1 = 4n^2 - 4n = 4n(n - 1).$$

But this resulting expression is indeed a multiple of 8 for any natural number n. For the factor 4 indicates that the number $4n(n - 1)$ is a multiple of 4. Furthermore, since n and $n - 1$ are consecutive integers, one of them must be even; hence, our product contains without fail still another factor 2. The number $4n(n - 1)$, therefore, is always a multiple of 8, which was to be proved.

From these examples we can see that there are two fundamentally distinct ways in which we gain knowledge of the world that surrounds us, of its objects, phenomena, and natural laws:

The first is that on the basis of a large number of observations and experiments on objects and phenomena we discover general laws. In the above examples, on the basis of observations men discovered the relation between the shape of a body and its shadow; numerous observations and experiments established the electromagnetic nature of light; tests carried out on the squares of odd numbers lead us to assert a certain property of such squares diminished by one. This method, drawing general conclusions from the observation of numerous particular cases, is called *induction* (from the Latin word "inductio"). Particular cases lead us to the idea of the existence of general laws.

A second method is used when, already knowing certain general laws, we apply this knowledge to particular cases. This method is called *deduction* (from the Latin word "deductio"). In the last example we applied the general laws of arithmetic to a particular case, thus proving a certain property of odd numbers. This last example also shows us that induction and deduction cannot be divorced from each other.

It is the combination of induction and deduction which characterizes the *scientific method*. In fact, in the course of every proof both these methods are involved. When seeking arguments for the proof of a proposition, we turn to experiments, observations, and facts, or else to already proved propositions. On the basis of such knowledge we then draw our conclusions regarding the truth or falsity of the proposition in question.

6. APPLICATION TO GEOMETRY

Geometry developed as the spatial properties of the material world were studied. By "spatial properties" we mean those relating to the shape, size, and relative position of objects. Of course, the importance of knowing such properties arises from our practical needs. We have to measure lengths, areas, and volumes in order to construct machines, buildings, roads, canals, etc. Naturally, man's first knowledge of geometry was obtained by the inductive method from a very large number of observations and experiments.

However, as the body of geometric knowledge grew, it was discovered that many truths could be obtained from others by means of deduction without resorting to observations or experiments. This idea occurred long ago to the geometers of ancient Greece, who began to develop a system of geometry in which the whole body of geometric truths known to them was deduced from a comparatively small number of fundamental propositions. Three hundred years before our era the Greek geometer Euclid of Alexandria developed the most extensive system of geometry of his time. In it he singled out certain propositions which he accepted without proof—the so-called "axioms" (from the Greek word "axios," which means "worthy, deserving confidence"). All other propositions, the truth of which is derived by means of proofs, came to be called "theorems" (from the Greek word "theoreo," which means "I think over" or "I meditate on").[1]

Euclid's system of geometry has endured for many centuries, and even in many schools today the presentation of geometry is essentially that of Euclid. We start with a comparatively small number of axioms, accepted without proof, and then derive all other propositions from the axioms by means of deductive reasoning. In this sense geometry is considered to be fundamentally a deductive science.

As a matter of fact, at the present time the work of many geometers is again concerned with the axioms of geometry and directed toward discovering all axioms necessary for the construction of a system of geometry, and, in so far as possible, toward reducing their number. Work along these lines was begun during the last century, and while a great deal has already been accomplished, this work is still not complete at the present time.

Summarizing our discussion, we can answer the question of what proof is in geometry by saying that *a proof is a chain of deductions through which the truth of the proposition to be proved is derived from axioms and previously established propositions.*

[1] *Editor's note.* For an additional discussion of the axiomatic method, see Raymond L. Wilder, *Introduction to the Foundations of Mathematics* (New York: Wiley, 1952), Chap. I–II.

2. Why Are Proofs Necessary?

7. THE LAW OF SUFFICIENT REASON

The need for proof is a consequence of one of the fundamental laws of logic (logic is the science of the laws of correct reasoning) —*the law of sufficient reason.* This law demands that any assertion we make should be well founded, that is, should be accompanied by sufficiently strong arguments supporting its truth, its agreement with the facts and with reality. Such arguments may be based either on references to possible verification through observation and experiment or on correct reasoning based on systematic deductions.

In mathematics we are concerned for the most part with arguments of the latter type.

8. DANGERS OF "OBVIOUSNESS"

The question sometimes arises as to whether it is necessary to give a proof when the proposition to be proved seems sufficiently clear and evident without it.

The Hindu mathematicians during the Middle Ages appear at first glance to have taken the point of view of omitting the proof. Frequently, they did not prove geometric propositions explicitly; instead they drew a clear figure illustrating them and then wrote above it the single word "Behold!" For example, in the book *Lilavati,* by the Hindu mathematician Bhaskara Acârya (born 1114 A.D.), the Pythagorean theorem is presented as in Fig. 2. From these two diagrams the student is to "discover" that the sum of the areas of the squares constructed on the legs of a right triangle is equal to the area of the square constructed on the hypotenuse.

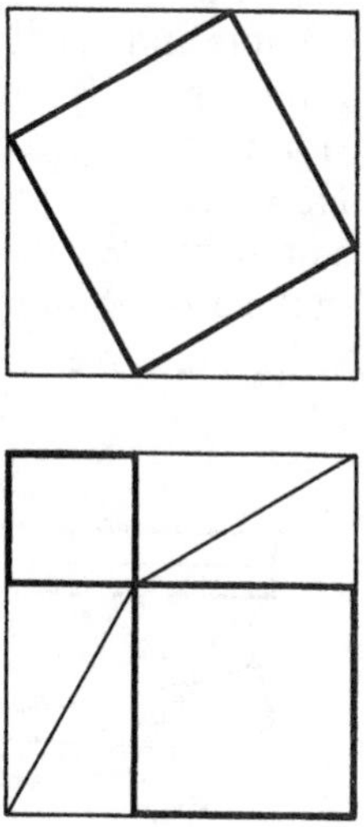

Fig. 2

Can we say that there is no proof in this case? Certainly not! If the student merely looked at the diagram, without reasoning, he would hardly come to any conclusion at all. However, the author assumes that the reader will not only look, but will think as well. He will then understand that the diagram shows two squares of equal size, that is, which have equal areas. The first square consists of four equal right triangles and a square constructed on their hypotenuses, while the second square consists of four of the same right triangles and two squares constructed on their legs. Now we have to reason that if from equals (the areas of the two equal large squares) we subtract equals (the areas of the four right triangles), there remain equals—in the first case a square constructed on the hypotenuse, in the second case two squares constructed on the legs. As we see, the conclusion here rests not merely on "obviousness," but on thinking and reasoning as well.

But perhaps there exist theorems in geometry which are actually so obvious that we could accept them without proof. As to that, let us first point out that, quite generally, we cannot rely on "obviousness" in an exact science. For the concept "obvious" is very vague; what seems obvious to one person may be quite doubtful to another. We need only remember how differently the same event is sometimes described by different observers, and how it is often difficult to determine the truth from the "testimony of witnesses."

Let me give you an interesting geometric example of how we may be deceived by apparent obviousness. I take a piece of paper and draw on it a continuous closed line. Then I take a pair of scissors and cut along this line. What happens to the piece of paper after the cut is completed? Most of us will answer without hesitation that the piece of paper will fall apart in two separate pieces. However, *this may not be the case.* Let us perform the experiment as follows: We take a strip of paper and glue the ends together to make a ring after having first turned over one of the ends of the strip. This gives us a "Möbius strip" (Fig. 3). (Möbius, 1790–1868,

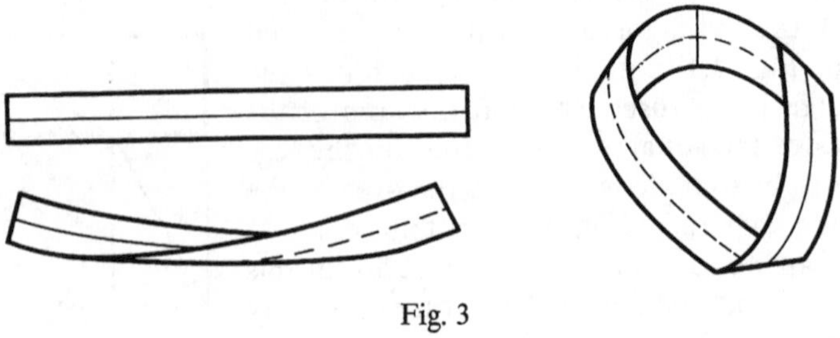

Fig. 3

was a German mathematician who studied surfaces of this sort.) If we now cut this strip along the closed line formed by the broken and solid lines shown in Fig. 3, we shall find that the paper will *not* fall into two separate pieces. Instead, we shall find ourselves holding a single, narrower strip in our hands. Occurrences such as this should make us careful about trusting conclusions based on "obviousness."

Let us look into this question more carefully. To the geometry student mentioned in section 1, it appeared strange that the teacher should first draw two apparently congruent triangles and then prove that they actually were congruent. The situation, however, was quite different. *The teacher did not draw two congruent triangles at all.* Instead, having drawn triangle ABC (Fig. 4), she said that the other triangle $A'B'C'$ was drawn in such a manner that $A'B' = AB$, $B'C' = BC$, and $\angle B' = \angle B$. This meant that we did not know whether $\angle A'$ and $\angle A$, $\angle C'$ and $\angle C$, and sides $A'C'$ and AC were respectively equal, for she did not set out to construct angles A' and C' respectively equal to the angles A and C, nor did she make the side $A'C'$ equal to the side AC.

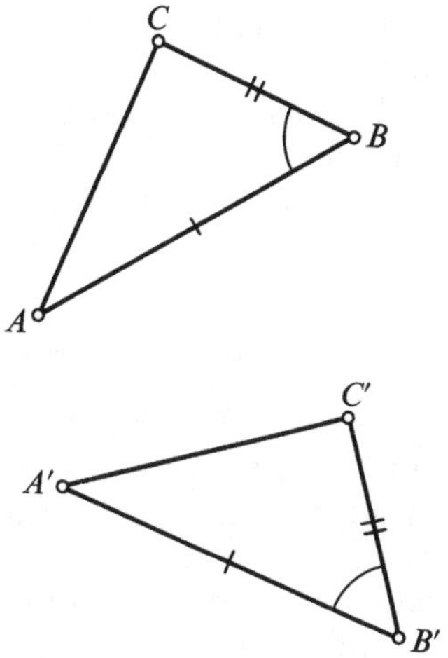

Fig. 4

Thus, we had to *deduce* the congruence of the triangles from the conditions $A'B' = AB$ and $B'C' = BC$ and $\angle B' = \angle B$; we had to deduce the equality of their remaining parts. This, of course, demanded reasoning, that is, proof. It is also easy to show that the congruence of triangles, on the basis of the equality of three pairs of their corresponding parts, is far from being as "obvious" as it might appear at first glance. For instance, let two sides of one triangle be respectively equal to two sides of the other, and also let an angle of the one be equal to the corresponding angle of the other. But let these angles be opposite a pair of corresponding sides which we are assuming equal, rather than included between the pairs of equal corresponding sides, which you recognize as one of the well-known conditions for the congruence of triangles. For example, in $\triangle ABC$ and $\triangle A'B'C'$ let $A'B' = AB$, $B'C' = BC$, and $\angle A' = \angle A$. What can we say about such triangles? By analogy with a well-known theorem on the congruence of triangles we might expect that these triangles would also be congruent. But Fig. 5 shows clearly that triangles ABC and $A'B'C'$, although constructed so as to satisfy the conditions $A'B' = AB$, $B'C' = BC$, and $\angle A' = \angle A$, are not congruent at all.

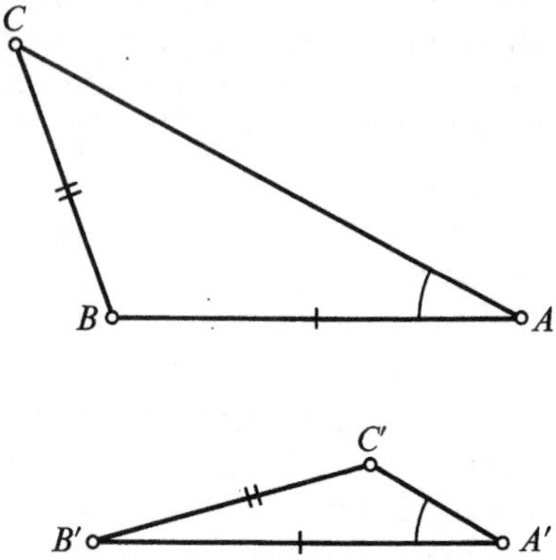

Fig. 5

Examples of this sort should make us careful in our judgments. They show that only a correctly constructed proof can guarantee the truth of the propositions we are trying to establish.

9. DANGERS OF PARTICULAR CASES

Now let us take a look at the theorem about the exterior angle of a triangle, which troubled my young neighbor Tolya (section 2). In the diagram shown in the standard textbook the exterior angle was indeed obtuse, while the nonadjacent interior angles were acute, which could readily be seen without any measurement. But does it follow from this that the theorem does not require a proof? It does not follow! For the theorem is not concerned with the particular triangle shown in the book, or on a piece of paper, or on the blackboard, but with *any* triangle whatsoever, even triangles which are very unlike the triangle in the textbook.

For example, let us imagine that point A moves away from point C in a straight line. Then triangle ABC will eventually take on a shape in which the angle at B is also obtuse (Fig. 6). If point A moves away from C about 10 meters, an ordinary protractor will no longer be able to detect the difference between the interior angle B and the exterior angle at C. And if A moves away from C to a distance equal to that between the earth and the sun, then not even the most exact modern instrument for measuring angles would be able to detect the difference between these angles. From this it

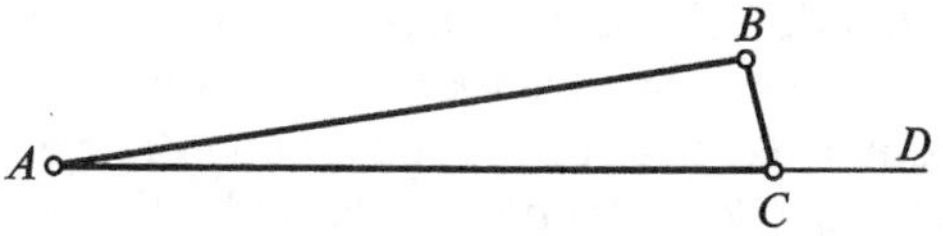

Fig. 6

is clear that there is nothing "obvious" about this theorem either. In fact, a rigorous proof of the theorem does not depend on the chance appearance of the triangle in the diagram, but demonstrates that the theorem about the exterior angle is valid for any triangle whatsoever, regardless of the relative length of its sides. Therefore, even in those cases where the difference between the interior and exterior angles is so small that it escapes detection by our measuring instruments, we are still certain that this difference exists. For we prove that in all cases the exterior angle of a triangle is greater than a nonadjacent interior angle.

In this connection let us turn our attention for a moment to the role of diagrams in the proof of geometric theorems. We should understand that the diagram is only an *aid* in the proof of the theorem, that it is only an example, only a particular case of a whole class of geometric figures for which the theorem in question is proved. Therefore, it is very important to distinguish the general, essential features of the figure from particular and accidental ones in a given diagram. For example, in the diagram for the theorem about the exterior angle of a triangle that is given in the textbook, it is by chance that the exterior angle is obtuse and the interior ones acute. Evidently, we must not rely on such chance facts in proving properties that are general for all triangles.

It is this aspect of geometric proof which makes it so necessary: the fact that it establishes the properties of spatial figures in all their generality. If the proof is reasoned correctly and is based on correct propositions, it will establish with absolute certainty the truth of the proposition being proved. For example, the proof of the Pythagorean theorem is valid for triangles of any dimensions, whether the length of the sides be a few millimeters or millions of kilometers.

10. GEOMETRY AS A SCIENTIFIC SYSTEM

Finally, there is still another very important reason why proof is indispensable. This is the fact that geometry is not a chance collection of truths describing the spatial properties of bodies, but a *scientific system,* based on strict laws. In this system each theorem is logically connected with the propositions previously established, and it is this connection which is disclosed by the proof. For example, the familiar theorem that the sum of the interior angles of a triangle is equal to 180° is proved on the basis of the properties of parallel lines. This reveals the immediate connection between the theory of parallel lines and the properties of the sums of the interior angles of polygons. Likewise, the entire theory of similarity of figures is found to be based on properties of parallel lines.

Thus, each geometric theorem is connected by a whole chain of deductions with previously proved theorems, and these with theorems that have been proved still earlier, and so on, and the chains of these deductions continue until we finally reach the basic definitions and axioms which are the foundation of the whole science of geometry. Such a chain of connections can be traced by taking any geometric theorem and considering all the propositions on which it is based.

11. SUMMARY

Thus, summing up what we have said about the necessity of proof, we may state the following:

(*a*) In geometry only a small number of fundamental propositions—axioms—are accepted without proof. The remaining propositions—theorems—are proved from these axioms by constructing a series of deductions.

(*b*) Proofs are needed because of a fundamental law of thinking—the law of sufficient reason, according to which we demand a rigorous foundation for the truth of our assertions.

(*c*) A correctly constructed proof relies only on axioms and previously proved propositions, not on "obviousness."[1]

(*d*) We also need proof to establish the generality of the proposition in question, that is, its applicability to all particular cases.

(*e*) Finally, by means of proofs, geometry becomes an orderly *system* of scientific knowledge, in which the connections between the different properties of spatial forms are disclosed.

[1] Many propositions of science which were once considered irrefutable because they were "obvious" have turned out to be false. In any science every proposition must be rigorously proved.

3. How Should a Proof Be Constructed?

12. CORRECT REASONING

We now move on to the question of the specifications a proof must meet so that we can call it *correct*. Notice first that every proof consists of a series of deductions. Therefore, the correctness or incorrectness of the proof depends on the truth or falsity of the deductions that enter into it.

As we saw above, deductive reasoning consists in the application of some general law to a given particular case. In order to prevent errors in our reasoning, we need to be familiar with certain methods by means of which the relations between concepts (in particular, geometric concepts) can be represented.

For example, let us assume that we have made the following deduction:

1) All rectangles have two equal diagonals.
2) All squares are rectangles.
3) Conclusion: All squares have two equal diagonals.

Now what do we have in this case? The first proposition states a general law, namely, that all rectangles, that is, the *whole set* of geometric figures which are called rectangles belong to the set of quadrilaterals which have equal diagonals. The second proposition asserts that the whole set of squares is a part of the set of rectangles. From this we draw the conclusion that the whole set of squares is a part of the set of quadrilaterals which have equal diagonals. Let us express this deduction in a more general form. We designate the most extensive of these sets (quadrilaterals which have equal diagonals) by the letter P, the intermediate set (rectangles) by the letter M, and the smallest set (squares) by the letter S. Then our deduction looks schematically like this:

1) All M are P.
2) All S are M.
3) Conclusion: All S are P.

It is helpful to depict this relationship by a diagram. Let us represent the largest set *P* by a large circle (Fig. 7), the set *M* by a smaller circle which lies entirely inside the first circle, and the set *S* by a still smaller circle which lies inside the second circle. There is no doubt, with the circles in these positions, that circle *S* lies wholly inside circle *P*. This representation of the relationships between concepts was proposed by the great Swiss mathematician Leonhard Euler (1707–1783).

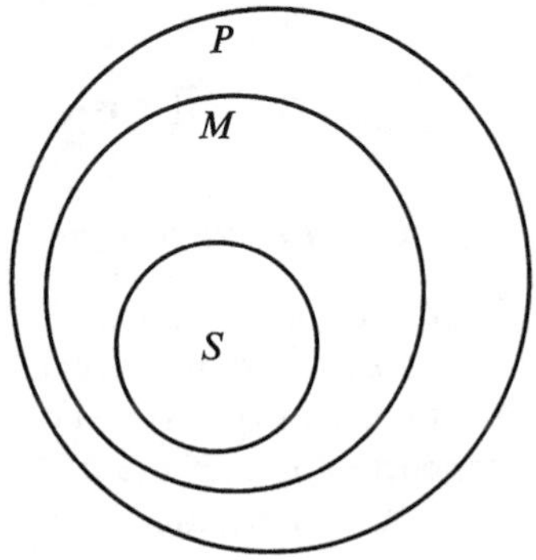

Fig. 7

By means of similar schemes we can represent other types of deductions. Let us look at one which has a negative conclusion:

1) No quadrilateral in which the sum of the opposite angles is different from 180° can be inscribed in a circle.
2) In an oblique parallelogram the sum of the opposite angles is not equal to 180°.
3) Conclusion: No oblique parallelogram can be inscribed in a circle.

Let us designate the set of quadrilaterals which can be inscribed in a circle by the letter *P*, the set of quadrilaterals the sum of whose opposite angles is not equal to 180° by the letter *M*, the set of oblique parallelograms by the letter *S*. Then our deduction can be set up according to the following scheme:

1) No *M* is *P*.
2) All *S* are *M*.
3) Conclusion: No *S* is *P*.

This relationship may also be represented by means of Euler circles (Fig. 8).

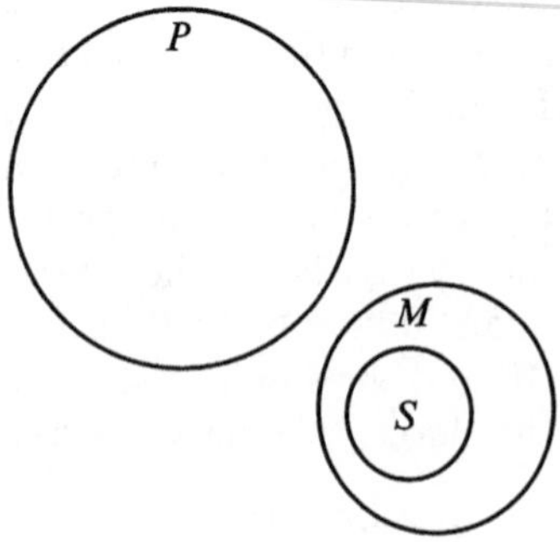

Fig. 8

The great majority of deductions in geometry are made according to one of these two preceding schemes. This method of representing the relations between geometric concepts greatly facilitates the understanding of the structure of deduction and helps to reveal mistakes in incorrect deductions.

13. INCORRECT REASONING

As an example let us analyze the incorrect reasoning of the student mentioned in section 3. His deductive argument was as follows:

1) In all quadrilaterals circumscribed about a circle the sums of the opposite sides are equal.
2) In the given trapezoid the sums of the opposite sides are equal.
3) Conclusion: The given trapezoid can be circumscribed about a circle.

Designating the set of quadrilaterals which can be circumscribed about circles by P, the set of quadrilaterals in which the sums of the opposite sides are equal by M, and the set of trapezoids in which the sum of the bases is equal to the sum of the lateral sides by S, we can express the boy's reasoning in terms of the following scheme:

1) All P are M.
2) All S are M.
3) Conclusion: All S are P.

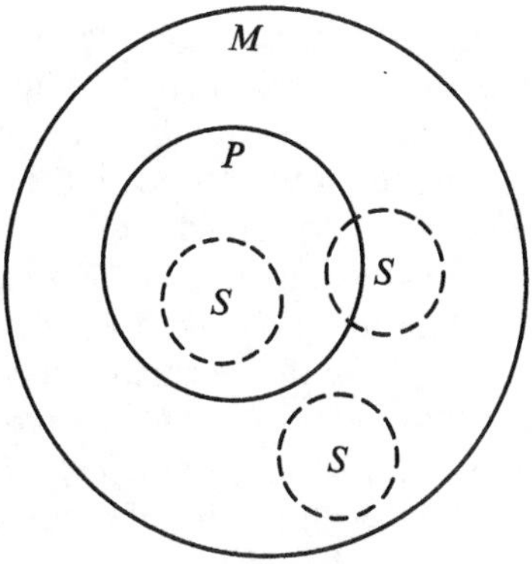

Fig. 9

This conclusion is clearly *incorrect*. Indeed, if we diagram the relationships between the sets by means of Euler circles (Fig. 9), we see that both P and S are inside M; but this does not permit any conclusion about the relationship between S and P.

14. CONVERSE THEOREMS

What is at the heart of the kind of faulty reasoning in section 13? It is the confusion of a "direct" theorem with its "converse." The direct theorem states: "In any quadrilateral in which a circle can be inscribed, the sums of the opposite sides are equal." The student without noticing it had used the converse of this theorem: "A circle can be inscribed in any quadrilateral in which the sums of opposite sides are equal." This converse theorem is not proved in the standard textbook, although it can be proved as we shall show below. If this theorem had been proved, the student could have made a correct deduction as follows:

1) In any quadrilateral in which the sums of the opposite sides are equal, a circle can be inscribed.
2) In the given trapezoid the sum of the bases is equal to the sum of the lateral sides.
3) Conclusion: It is possible to inscribe a circle in the given trapezoid.

This deduction is, indeed, correct, as can be seen from Fig. 7.

1) All *M* are *P*.
2) All *S* are *M*.
3) Conclusion: All *S* are *P*.

In other words, the student's error consisted in using in his proof the direct theorem, when he should have used the converse. In making the distinction between "direct" and "converse" theorems, it must be understood that in such a pair of propositions neither one, as such, is "direct" or "converse," but either one of the two is the converse of the other. We may, of course, designate as the "direct" theorem the one which was proved first.

We now prove this important converse theorem.

THEOREM. *It is possible to inscribe a circle in any quadrilateral in which the sums of the opposite sides are equal.*

We first note that if a circle can be inscribed in a quadrilateral, then its center is equidistant from all four sides. And since the locus of points which are equidistant from the sides of an angle is the bisector of the angle, the center of the inscribed circle must lie on the bisector of each interior angle. Thus, the center of the inscribed circle is the point of intersection of the four bisectors of the interior angles of the quadrilateral.

It is easy to show that if the bisectors of any three of the angles of a quadrilateral intersect in some point, then the bisector of the fourth angle also passes through this point, which means that this point will be equidistant from all four sides and, hence, be the center of an inscribed circle. (This can be proved by the same reasoning used in proving the theorem on inscribing a circle in a triangle. Hence, we leave this part of the proof to the reader.)

Now let us proceed with the essential part of the proof. We are given a quadrilateral $ABCD$ (Fig. 10) for which the following relation holds:

$$AB + CD = BC + AD. \qquad (1)$$

We exclude the case where the quadrilateral is a rhombus, since in a rhombus the diagonals are bisectors of the interior angles and, therefore, their point of intersection is at once seen to be the center of the inscribed circle; it is always possible to inscribe a circle in a rhombus. Let us suppose, therefore, that in this quadrilateral there are two unequal adjacent sides, say $AB > BC$. Then, from equation (1), we obtain $CD < AD$. On AB we now lay off the segment $BE = BC$ and obtain the isosceles triangle BCE. Next, on AD we lay off the segment $DF = CD$, obtaining the isosceles triangle CDF.

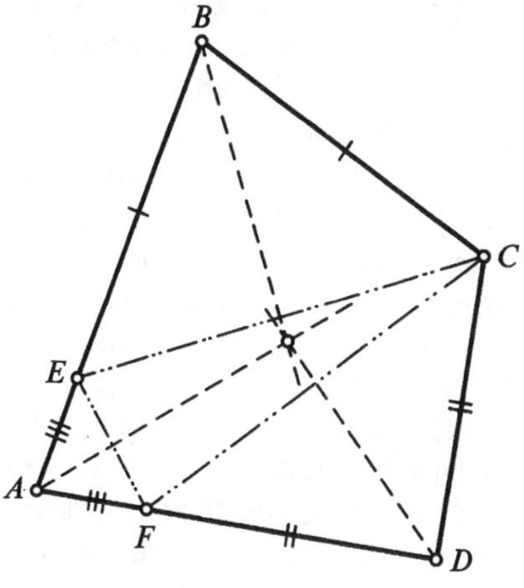

Fig. 10

We shall next show that $\triangle AEF$ is also isosceles. Subtracting $BC + CD$ from both sides of equation (1), we obtain $AB - BC = AD - CD$. But $AB - BC = AE$, and $AD - CD = AF$. Hence, $AE = AF$; that is, $\triangle AEF$ is isosceles. In these three isosceles triangles we next draw bisectors of the vertex angles, that is, the bisectors of $\angle B$, $\angle D$, and $\angle A$. These three bisectors are respectively perpendicular to the bases CE, CF, and EF and also bisect these bases. Consequently, being the perpendicular bisectors of the sides of triangle CEF, they intersect in one point. This proves that three angle bisectors of our quadrilateral intersect in one point, which, as was pointed out above, is then the center of the inscribed circle.

15. DISTINGUISHING BETWEEN DIRECT AND CONVERSE THEOREMS

The mistake of quoting a direct theorem where, in fact, its converse is being used occurs frequently, and we have to guard carefully against it. For example, if a student is asked what kind of triangle has sides respectively 3, 4, and 5 units long, he might answer a right triangle because $3^2 + 4^2 = 5^2$ and, according to the Pythagorean theorem, in any right triangle the sum of the squares of the legs is equal to the square of the hypotenuse. In fact, he should quote the converse of the Pythagorean theorem: "Any triangle in which the sum of the squares of two sides is equal to the square of the third side is a right triangle." Although this converse theorem is proved in the standard textbook, often not enough attention is paid to it, and the mistake just mentioned is made.

In this connection it is useful to examine the conditions under which a theorem and its converse are both true. We already know examples where this occurs, but we can give just as many examples where one is true and the other is false. For example, a "direct" theorem correctly asserts that all vertical angles are equal, whereas its converse theorem, that all angles which are equal to each other are vertical angles, is, of course, false.

We can use Euler circles to represent the relation between a theorem and its converse. Let a theorem be stated in the form "all *S* are *P*." (For example: "All pairs of vertical angles are pairs of equal angles.") Then the converse asserts "all *P* are *S*." ("All pairs of equal angles are pairs of vertical angles.") By representing the direct theorem by means of Euler circles, we obtain Fig. 11, which shows the set *S* to be a part, or "subset," of the set *P*. The only valid conclusion which can be drawn from this is that "some *P* are *S*": "Some pairs of angles which are equal to each other are pairs of vertical angles."[1]

Under what conditions will the proposition "all *S* are *P*" and the proposition "all *P* are *S*" both be true? It is obvious that this

[1] *Editor's note.* In this example we have deduced from "all *S* are *P*" that "some *P* are *S*." If we are absolutely rigorous, we cannot even admit this deduction. For example, suppose that we have already proved that "if a triangle has two right angles, then it is isosceles." Letting *S* be the class of all triangles with two right angles, and letting *P* be the class of all isosceles triangles, our statement reads "all *S* are *P*." However, we cannot conclude that "some *P* are *S*"; in fact, no triangle has two right angles. In the example on pairs of vertical angles, the conclusion is true because there do exist pairs of vertical angles. The deduction of "some *P* are *S*" from "all *S* are *P*" is perfectly valid if we know that, at the same time, there are some *S*.

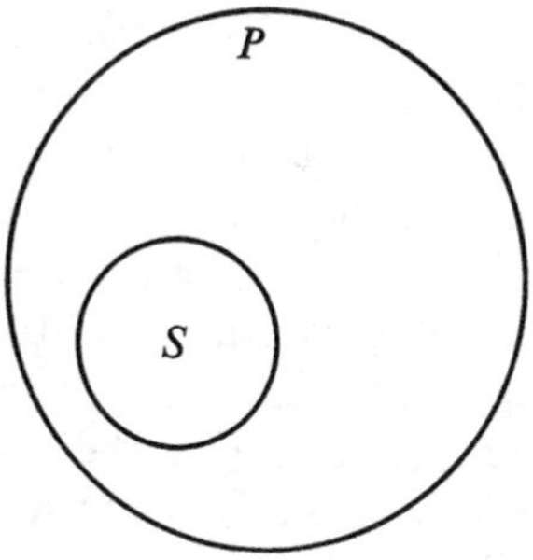

Fig. 11

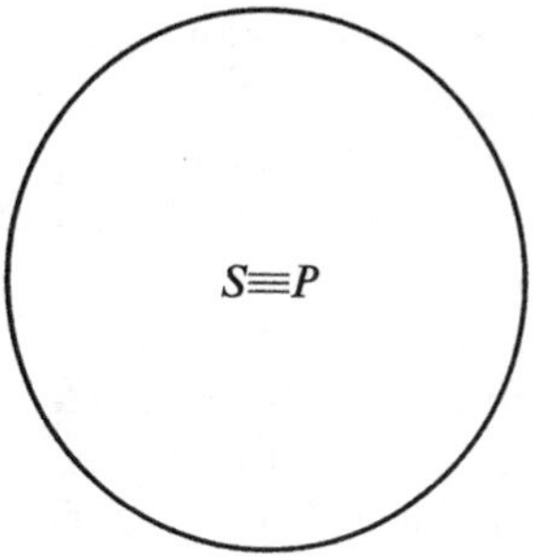

Fig. 12

happens if and only if the sets S and P are *identical* ($S \equiv P$). In this case the circle which represents S coincides with the circle which represents P (Fig. 12). For example, the theorem "all isosceles triangles have equal base angles" and its converse "all triangles that have equal base angles are isosceles" are both true. This is because the set of isosceles triangles and the set of triangles with equal base angles are the same set. Likewise, the set of right triangles and the set of triangles in which the square of one of the sides is equal to the sum of the squares of the other two sides coincide.

Our student in section 3 was lucky to get the right solution despite the fact that he used the direct theorem instead of the converse. But this was possible only because the set of quadrilaterals in which a circle can be inscribed coincides with the set of quadrilaterals in which the sums of the opposite sides are equal. (In this case both "all P are M" and "all M are P" are true—see the proof in section 14.)

This discussion also shows clearly that the truth of a theorem never implies the truth of its converse, which *always requires a separate proof.*

16. CONDITIONAL AND CATEGORICAL STATEMENTS

It sometimes happens that a theorem and its converse are not stated as sentences of the form "all S are P" and "all P are S." This is the case when the theorems are expressed in the form of so-called "conditional" statements: "If A is B, then C is D." For example: "If a quadrilateral is circumscribed about a circle, then the sums of its opposite sides are equal." The first part of the proposition, "if A is B," is called the *condition*, or *hypothesis*, of the theorem; the second part, "then C is D," is called the *conclusion.* The converse is obtained by turning the conclusion into the hypothesis, and the hypothesis into the conclusion. In many instances the formulation of a theorem as a conditional statement is more natural than as a "categorical" statement, "all S are P." However, this distinction is immaterial. In fact, any conditional statement can be changed into the categorical form, and vice versa. For example, take the following theorem expressed in the conditional form: "If two parallel lines are intersected by a third line, then alternate interior angles are equal." Expressed in the categorical form it reads: "Lines which are parallel are also lines which form equal alternate interior angles when intersected by a third line." Thus, our preceding discussion applies equally well to theorems expressed in the conditional form. In this example the direct and converse theorems are simultaneously true because the sets of objects satisfying the corresponding conditions coincide. Thus, in the above example the set of "parallel lines" is the same as the set of "straight lines which, if intersected by a third, form equal alternate interior angles."

17. AVOIDING PARTICULAR CASES

Let us next examine some other types of errors which occur in proofs. Often a proof is wrong because it treats accidental, particular cases without dealing with the general situation. This was the mistake in the reasoning of my young neighbor Tolya (section 2), who, in order to prove a general theorem about exterior angles of any triangle, considered only an acute triangle, in which, of course, all exterior angles are obtuse and all interior ones acute.

Let us take another example of an erroneous proof of this kind, but this time one in which the error is much less obvious. We gave above an example of two triangles which were not congruent (see Fig. 5), even though two sides and the angle opposite one of them in one triangle were respectively equal to the corresponding parts of another triangle. Let us now give a "proof" that two triangles which satisfy these very conditions are congruent. This proof is also interesting because it is similar to the (correct) proof in the standard textbook of the theorem stating that two triangles are congruent if all three of their corresponding sides are equal.

In $\triangle ABC$ and $\triangle A'B'C'$ (Fig. 13) let $AB = A'B'$, $AC = A'C'$, and $\angle C = \angle C'$. To show that the two triangles are congruent, we

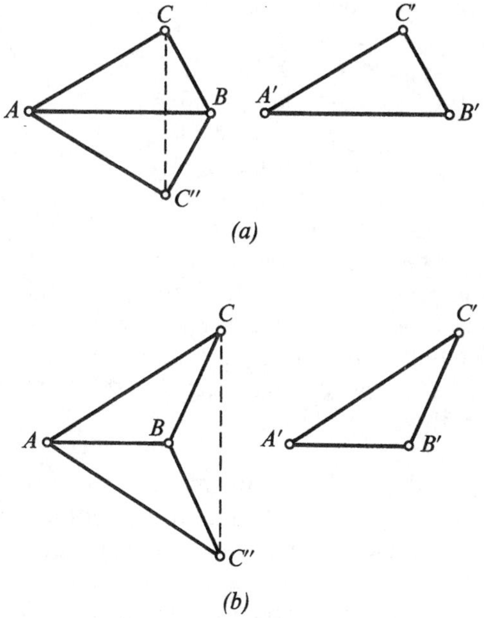

Fig. 13

place $\triangle A'B'C'$ on $\triangle ABC$ in such a way that side $A'B'$ coincides with side AB and point C' takes the position C''. We connect points C and C''. Let us assume that the segment CC'' intersects side AB between points A and B (Fig. 13*a*). According to the given conditions, $\triangle ACC''$ is isosceles ($AC = AC''$); hence, $\angle ACC'' = \angle AC''C$. Also, since $\angle C = \angle C''$, subtracting equal angles from equal angles, we get $\angle BCC'' = \angle BC''C$, which means that $\triangle CBC''$ is also isosceles; therefore, $BC = BC''$. This means that all three sides of $\triangle ABC$ are respectively equal to those of $\triangle A'B'C'$. Hence, $\triangle ABC \cong \triangle A'B'C'$.

Even if segment CC'' intersects the straight line AB beyond the segment AB (Fig. 13*b*), the proof will remain valid. As before, $\triangle ACC''$ is isosceles; hence, $\angle ACC'' = \angle AC''C$. Since $\angle C = \angle C''$, we again find that $\angle BCC'' = \angle BC''C$. Thus $\triangle BCC''$ is isosceles ($BC = BC''$), and again we have proved $\triangle ABC \cong A'B'C'$.

We seem to have given a complete proof, having accounted for all possible cases. In fact, however, we have overlooked one more possibility, namely, the case when segment CC'' passes through the end of segment AB. In that case segment CC'' passes through point B (Fig. 14). It is easy to see that in this case our reasoning is no longer applicable, and the two triangles may not be congruent, as shown in Fig. 14.

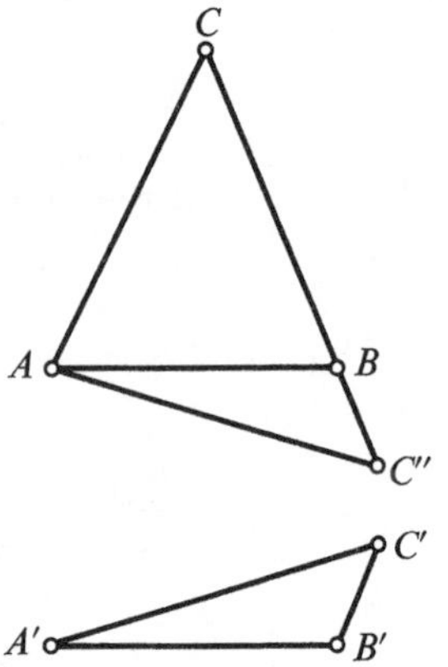

Fig. 14

Another instructive example of mistakes of this type can easily occur in theorems about prisms. One such theorem asserts, "The area of the lateral surface of a prism is equal to the product of the perimeter of a perpendicular section and the length of a lateral edge." Another states, "The volume of any prism is equal to that of a right prism whose base is a perpendicular section of the given prism and whose altitude is equal to a lateral edge of the given prism." The statements of these theorems, much less the proof of them, make no sense unless we consider only cases of prisms *with* perpendicular sections. We must ignore a whole class of prisms in which it is *impossible to make a perpendicular section that will intersect all of the lateral edges,* that is, highly oblique prisms of very small altitude (Fig. 15). In such a prism a section perpendicular to one of the lateral edges will not intersect the remaining edges, and all the reasoning used in the proof of these propositions turns out to be inapplicable. This oversight occurred because of our deep-rooted habit of thinking of a prism as a bar of fairly large altitude, while flat, slab-like prisms are almost never shown on the blackboard or in textbooks. This example shows again with what caution we must base our arguments on the diagrams illustrating a proof. Every time we draw a diagram needed in the course of a proof, we must ask ourselves, "Is it possible to use this diagram in *all* cases?" If such a question had been asked during the proof of the above-

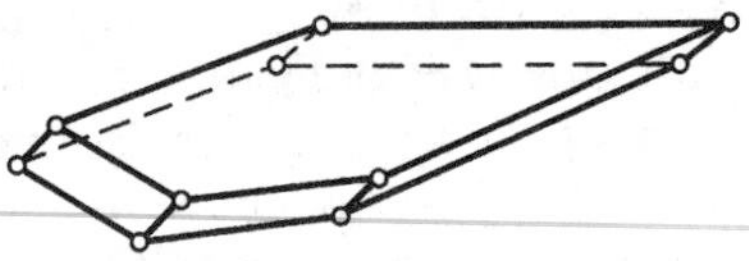

Fig. 15

mentioned propositions about an oblique prism, it would not have been difficult to find examples of prisms in which no perpendicular section could be constructed.

18. INCOMPLETE PROOFS

In the last two examples the mistake would be that, based on the usual conception of a prism, the proof would apply not to the general proposition to be proved, but only to special cases. We shall next give another example where the mistake is of a somewhat similar nature. This mistake is far more crucial and also more difficult to detect.

It concerns the proof of the existence of incommensurable segments as it is usually given in school courses in elementary geometry. Let us recall the method of reasoning in this proof. First, one defines a "common measure" of two line segments to be a line segment that can be laid off a whole number of times on each of the two given segments. One then proves that it can be laid off a whole number of times also on the sum and the difference of the two segments. Next we explain a method of finding the largest common measure of two segments—a method used even by Euclid. It consists in first laying off the smaller segment on the larger as many times as possible; if there is a remainder, we lay it off on the smaller segment; if there is again a remainder, we lay it off on the first remainder; and so on. If, finally, a remainder is obtained which can be laid off a whole number of times on the preceding remainder, then this final remainder is the largest common measure of the two segments. Segments which have a common measure are called *commensurable;* segments which have no common measure are called *incommensurable*. But do such incommensurable segments exist? To prove this we need only establish the existence of at least one pair of incommensurable segments. As an example of such a pair we usually take the side of a square and one of its diagonals. The proof, in accordance with Euclid, consists in first laying off the side on the diagonal; then laying off the remainder on the side, whereby a new remainder is obtained. This new remainder is then shown to be the side of a new square whose diagonal is the first remainder. This process, repeated on the new square, leads to a third square in which the third remainder is the side and the second remainder the diagonal. This process can be continued indefinitely without coming to an end. Hence, it is impossible to find a common measure for the side of a square and its diagonal. Now a further conclusion is drawn: the segments are incommensurable!

This conclusion, however, is not justified—at least not on the basis of what has been said so far. For we have only shown that no common measure can be found by Euclid's method, not that a common measure does not exist! If we fail to discover an object we are looking for by a certain method, this does not mean that the object does not exist. It might be found by some other method. We would not accept the argument: "Electrons cannot be seen with a microscope; therefore, they do not exist." Rather we would reply: "There are other methods by which we ascertain the existence of electrons."

We shall next show that additional argument is needed to make the proof of the existence of incommensurable segments conclusive. For this purpose we must prove the following proposition: *If Euclid's method of determining the greatest common measure of two segments can be indefinitely continued, then no common measure exists.*

For this proposition we shall now give an indirect proof. Let there be given two segments $\bar{a}$ and $\bar{b}$, and let $\bar{a} > \bar{b}$. (We are denoting segments by dashed letters, and numbers by undashed letters.) We now lay off $\bar{b}$ on $\bar{a}$ until the first remainder $\bar{r}_1$ is obtained; then we lay off $\bar{r}_1$ on $\bar{b}$ until the second remainder $\bar{r}_2$ is obtained; then $\bar{r}_2$ on $\bar{r}_1$; and so on. Let us assume that this process is unending, that is, that an infinite sequence of remainders $\bar{r}_1, \bar{r}_2, \bar{r}_3, \ldots$ is obtained. Each term in this sequence is smaller than the preceding one; that is, $\bar{a} > \bar{b} > \bar{r}_1 > \bar{r}_2 > \bar{r}_3 > \cdots$.

Now, if there existed a common measure $\bar{p}$ for $\bar{a}$ and $\bar{b}$, it would be possible to lay it off a whole number of times on $\bar{a}$, on $\bar{b}$, and on each of the remainders $\bar{r}_1, \bar{r}_2, \bar{r}_3, \ldots$. Suppose it can be laid off m times on $\bar{a}$, n times on $\bar{b}$, n_1 times on $\bar{r}_1$, n_2 times on $\bar{r}_2, \ldots, n_k$ times on $\bar{r}_k$, and so on. The numbers $m, n, n_1, n_2, n_3, \ldots$ all are positive integers, while the above inequality of the segments implies the corresponding inequality of these numbers:

$$m > n > n_1 > n_2 > n_3 > \cdots.$$

Since it is assumed that the sequence of segments continues indefinitely, the sequence of numbers $m, n, n_1, n_2, n_3, \ldots$ must also continue indefinitely. This, however, is impossible, since a decreasing sequence of positive integers cannot be infinite. We are, therefore, forced to give up the assumption that, in spite of the infinity of the sequence of remainders, there exists a common measure for the two segments $\bar{a}$ and $\bar{b}$; that is, we have proved that these segments are incommensurable. In the example of the side of a square and its diagonal the geometric argument had shown that the process of obtaining remainders is infinite. This fact, together with the proposition proved just now, establishes that the side of a square and its diagonal are incommensurable and, hence, that incommensurable segments exist.

Without this supplementary proposition the proof is incomplete.

19. CIRCULAR REASONING

A different kind of mistake occurs where in the course of the proof reference is made to a proposition which has not yet been proved. It even occasionally happens that a student uses in a proof the very proposition he is trying to prove. For example, one may occasionally hear a conversation like this between teacher and pupil: The teacher asks, "Why are these straight lines perpendicular?" The pupil answers, "Because the angle between them is a right angle." "But why is the angle a right angle?" "Because the straight lines are perpendicular."

A mistake of this sort is called "circular reasoning." It is but rarely met in such a glaring form; it is usually disguised. For example, a student is given the problem: "Prove that if two angle bisectors of a triangle are equal, then the triangle is isosceles." He may give this proof: "In $\triangle ABC$ the angle bisector AM is given equal to bisector BN (Fig. 16). Then $\triangle ABM$ is congruent to $\triangle ABN$, because $AM = BN$, AB is a common side, and being halves of the equal base angles, $\angle ABN = \angle BAM$. From the congruence of $\triangle ABM$ and $\triangle ABN$ it follows that $AN = BM$. Next, it follows that $\triangle ACM \cong \triangle BCN$, since $AM = BN$ and the corresponding angles adjacent to these sides are respectively equal. Therefore, $NC = MC$, which means that $AN + NC = BM + MC$; that is, $AC = BC$."

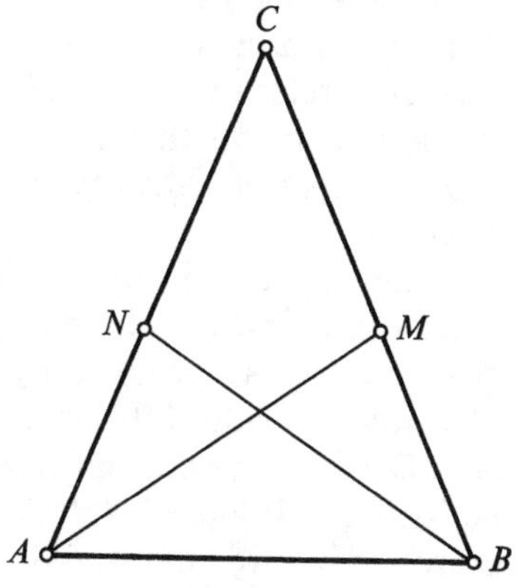

Fig. 16

The error in this proof consists in assuming the equality of the base angles of the triangle, since, in fact, the equality of these angles is a consequence of the triangle's being isosceles, which is the very proposition to be proved.

Again, there are "proofs" which rely on unproved propositions which are regarded as obvious, although they are not included among the axioms. Let us look at two examples of this kind. In studying the relative positions of a straight line and a circle, three cases are considered: 1) the distance of the straight line from the center of the circle is greater than the radius so that the straight line lies *outside* the circle; 2) if the distance of the straight line from the center is equal to the radius, then the straight line has one and only one point in common with the circle (tangent); 3) if the distance of the straight line from the center is less than the radius, then the straight line has exactly two points in common with the circle (secant).

The first two propositions are easily furnished with correct proofs; in the case of the third, the standard text states, "The straight line passes through a point lying inside the circle, and hence, *obviously* intersects the circle." As a matter of fact, this word "obviously" conceals a significant geometric proposition, namely: "Any straight line that passes through a point inside a circle intersects the circle." It is true that this proposition seems quite obvious, but we have already pointed out how vague and indefinite the concept "obvious" is. Therefore, this proposition must either be included among the axioms, or else it must be proved by means of other propositions.

As a second example we give an incorrect proof of the converse of a theorem mentioned earlier concerning quadrilaterals and inscribed circles (a proof which can even be found in some texts). We are to prove that *if a quadrilateral is such that the sums of opposite sides are equal, then a circle can be inscribed in the quadrilateral.*

The "proof" is as follows: "Given $AB + CD = BC + AD$ (Fig. 17), we draw a circle tangent to three sides of the given quadrilateral, say AB, BC, and CD. We have to show that this circle also touches side AD. Let us suppose that it does not touch side AD. Drawing the tangent AD_1 from point A, we obtain the circumscribed quadrilateral $ABCD_1$; it follows from the theorem whose converse we are now studying that $AB + CD_1 = BC + AD_1$. Subtracting this equation from the given equation, we get $CD - CD_1$

$= AD - AD_1$ or $DD_1 = AD - AD_1$. This, however, is impossible, since the difference between two sides of $\triangle ADD_1$ cannot be equal to the third side. Therefore, the circle tangent to sides AB, BC, and CD must also be tangent to AD."

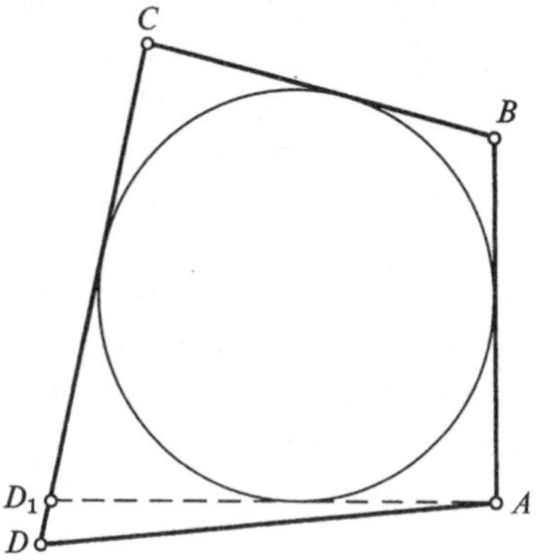

Fig. 17

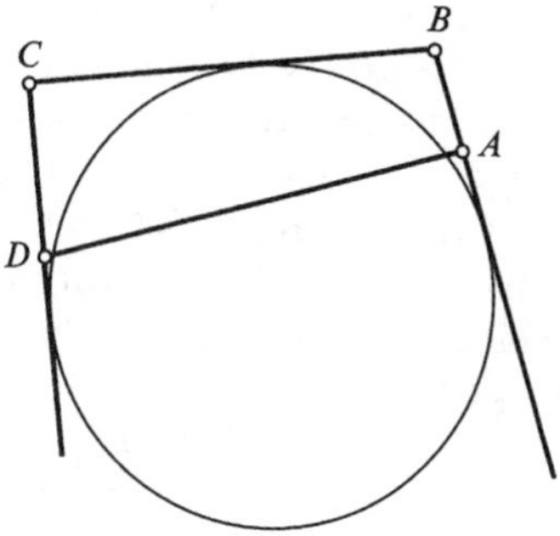

Fig. 18

The error in this proof consists in assuming, without proof, that *the point of tangency of the circle and side AB lies between points A and B*. When points A and D have the positions indicated in Fig. 18, then it is impossible to use the argument given above. As a matter of fact, it is quite possible to prove that the points of tangency must lie between A and B and between C and D, but this requires a rather lengthy argument.

20. REQUIREMENTS FOR A CORRECT PROOF

Thus, our question, "What requirements must a proof satisfy in order to be valid (that is, to guarantee the truth of the proposition to be proved)?" can be answered as follows:

(*a*) The proof must be based only on axioms or previously proved theorems.

(*b*) All deductions by which the proof is established must be carried out correctly.

(*c*) We must always keep in mind the purpose of the proof, which is to establish the truth of the proposition to be proved, and not substitute some other proposition in its place.

In view of the necessity of satisfying these requirements, the question of how we can find correct proofs arises.

21. HOW TO FIND A CORRECT PROOF

First of all, when we are asked to prove a geometric proposition, it is necessary to express precisely *the basic assertion* which is to be proved. Often this is not done clearly enough. For example, suppose the proposition is: "Prove that by joining successively the mid-points of the sides of a quadrilateral, we obtain a parallelogram." By what means can we prove that we get a parallelogram? We recall the *definition* of a parallelogram as a quadrilateral having its opposite sides parallel. This means that we must prove that pairs of the segments obtained are parallel.

After we have determined exactly what we are to prove, we must extract from the wording of the given theorem the conditions which are given and those which are essential for the proof. In the present example it is stated that we connect the *mid-points* of the sides of the quadrilateral. This means that points are taken which divide the sides into equal parts.

All this is next set down in the symbolic form customarily used in school, under the headings "Given" and "To prove." Thus, in our example, if we have quadrilateral $ABCD$ (Fig. 19) with M, N, P, Q as the mid-points of its sides, we can write our theorem as follows:

Given: The quadrilateral $ABCD$, in which $MA = MB$, $NB = NC, PC = PD, QD = QA$.

To prove: $MN \| PQ$, $MQ \| NP$.

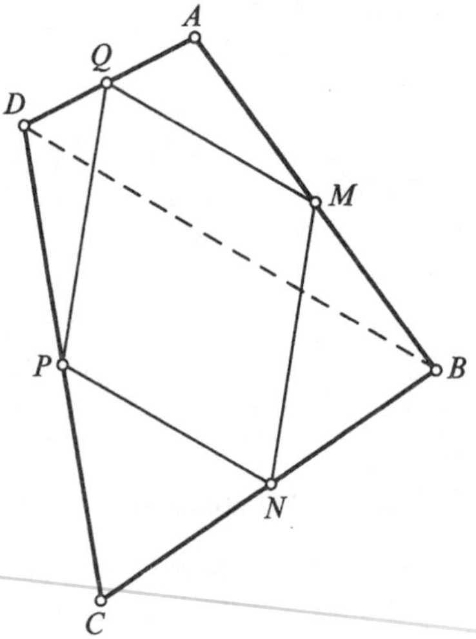

Fig. 19

After writing this down, we proceed to prove the theorem. In the proof we make use of axioms or previously established theorems and, of course, of those special relations which are "given" in the theorem. We have to find a chain of reasoning which will lead from these to the proposition to be proved.

22. ANALYSIS

How are we to select from all the many propositions the particular ones that will serve to prove our theorem? In this search it is best to start off from the proposition which is to be proved and to state the question thus: By deduction from what proposition can we get the proposition which we wish to prove? If such a proposition can be found, and if it is a consequence of the conditions and of previously proved theorems, then our problem is solved. If not, however, we must restate the same question for another proposition, and so on. In scientific reasoning such a train of thought is called *analysis*.

In the case of the quadrilateral which we were examining in section 21, we have to prove that certain segments are parallel. At the same time, we see that these segments join the mid-points of the sides of the quadrilateral. With this in mind we ask ourselves whether there is perhaps, among previously proved propositions, one concerning the parallelism of segments that join the mid-points of the sides of a polygon. One such proposition is the theorem which states that a segment joining the mid-points of two sides of a triangle is parallel to the third side and equal to half its length. In the figure which we are examining there are no such triangles, but we can at once construct one by drawing one of the diagonals, say, BD. It divides the figure into the two triangles ABD and BCD, in which the segments MQ and NP connect the mid-points of the two sides. Thus, $MQ \| BD$ and $NP \| BD$, and, hence, $NP \| MQ$. By drawing the second diagonal we prove, in the same way, that $MN \| PQ$.

Alternatively, we may not need the second construction, since from the first pair of triangles we know also that $MQ = \frac{1}{2} BD$ and $NP = \frac{1}{2} BD$; hence, $MQ = NP$. Thus, the opposite sides MQ and NP of the quadrilateral $MNPQ$ are not only parallel, but also equal. Then if we have already proved the proposition stating, "If a pair of opposite sides of a quadrilateral are parallel and equal, the quadrilateral is a parallelogram," we can consider our proof as complete.

As a second example let us take the well-known theorem concerning the sum of the interior angles of a triangle. In this case no special conditions are "given"; therefore, we only need to write down "to prove": In $\triangle ABC$ (Fig. 20) $\alpha + \beta + \gamma = 180°$. To prove our theorem, we must apparently add the three interior angles of the triangle. This addition is performed most expediently in the figure itself. At the vertex B (with angle β) we construct the angle $\gamma' = \gamma$ adjacent to β. Then line BD is parallel to AC because the alternate interior angles γ and γ' with BC are equal. By extending side AB beyond point B, we get $\angle DBE$, which we designate as α'. Then $\alpha' = \alpha$, because they are corresponding angles on the parallel lines AC and BD cut by line AB. Thus, we have $\alpha' + \beta + \gamma' = 180°$, since these angles together make up a straight angle. But $\alpha' = \alpha$ and $\gamma' = \gamma$, so that we obtain the relationship to be proved,

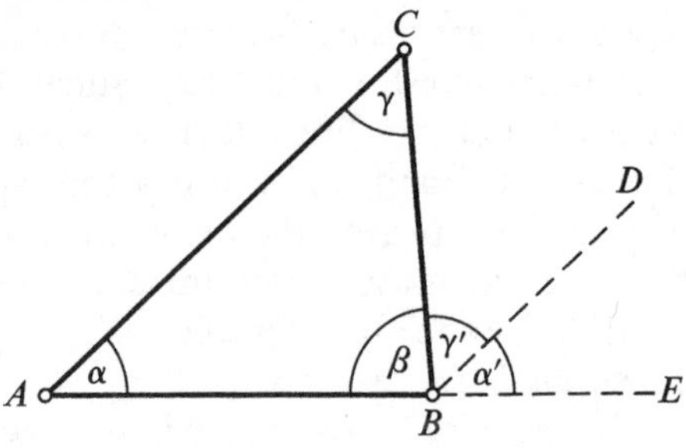

Fig. 20

$$\alpha + \beta + \gamma = 180°.$$

In both these examples it was easy enough to think of suitable relations to be used. But there are situations in which a whole chain of subsidiary propositions has to be mobilized, so that the analysis becomes quite long and complicated.

Let us take an example involving a more complex analysis. The following proposition is to be proved: *If a circle is circumscribed about a triangle, and if from an arbitrary point on the circle perpendiculars are dropped to the sides of the triangle, their points of intersection with the respective sides lie on a straight line* (*Simpson's line*).

Let us carry out the analysis. Let ABC be the given triangle (Fig. 21), M a point on the circumscribed circle, and N, P, Q the projections of this point on the sides BC, CA, and AB, respectively. We are to prove that N, P, and Q lie on a straight line. We begin by formulating the proposition to be proved, keeping in mind that the condition for points N, P, and Q to lie on a straight line is equivalent to saying that angle NPQ is a straight angle. Thus:

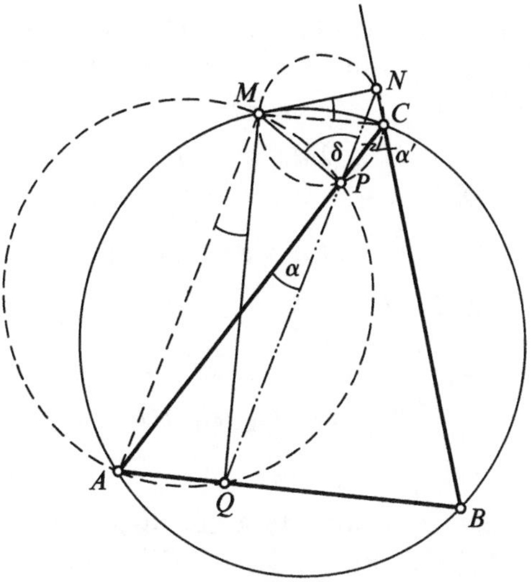

Fig. 21

Given: $MN \perp BC$, $MP \perp CA$, $MQ \perp AB$; point M lies on the circle circumscribed about $\triangle ABC$.

To prove: $\angle NPQ = 180°$.

If we look at angle NPQ, we see that it consists of $\angle MPN = \delta$, $\angle MPA = 90°$, and $\angle APQ = \alpha$. The proposition will, therefore, be proved if we succeed in proving that $\angle NPQ = \delta + 90° + \alpha = 180°$, or simply that $\alpha + \delta = 90°$. Let us examine $\angle CPN = \alpha'$. Since $\angle MPC = 90°$, it follows that $\alpha' + \delta = 90°$. Now if we can show that $\alpha' = \alpha$, the theorem will be proved. Let us try to establish this equality by examining some other angles determined by the given conditions. The angles APM and AQM are right angles; therefore, the circle constructed with AM as a diameter passes through points P and Q. By virtue of the properties of inscribed angles, $\angle AMQ = \angle APQ = \alpha$. In the same way, we construct a circle with MC as a diameter and see that it must pass through P and N, and, again by virtue of the properties of inscribed angles, $\angle CMN = \angle CPN = \alpha'$. We shall now try to prove that $\angle AMQ = \angle CMN$. To do this, let us note that $ABCM$ is an inscribed quadrilateral; therefore, the sum of its opposite angles is 180°:

$$\angle AMC + \angle B = 180°,$$

or

$$\angle AMQ + \angle QMC + \angle B = 180°. \tag{1}$$

On the other hand, in quadrilateral $BQMN$ the angles at points Q and N are right angles; therefore, the sum of the other two angles is equal to 180°:

$$\angle QMN + \angle B = 180°,$$

or

$$\angle QMC + \angle CMN + \angle B = 180°. \quad (2)$$

From equalities (1) and (2), we next obtain

$$\angle QMC + \angle CMN + \angle B = \angle AMQ + \angle QMC + \angle B;$$

hence,

$$\angle CMN = \angle AMQ,$$

that is, $\alpha' = \alpha$.

From this, as we have already seen, it follows that $\alpha + \delta = 90°$, $\alpha + \delta + 90° = 180°$, and, finally, $\angle NPQ = 180°$.

23. SYNTHESIS

To write out a proof, all the successive steps of the analysis would have to be taken in reverse order. In the last example of section 22, we would first prove that $\angle AMQ = \angle CMN$, and then we would establish the equalities

$$\angle AMQ = \angle APQ \quad \text{and} \quad \angle CMN = \angle CPN.$$

From the fact that $\angle CPA = \angle CPN + \angle MPN + 90° = 180°$, we would find that $\angle NPQ = \angle MPN + 90° + \angle APQ = 180°$, that is, that points N, P, and Q lie in a straight line.

This way of presenting a proof, which is the one ordinarily used in textbooks and in class, is in the reverse order of the analysis and is called *synthesis*. The proof of a theorem by the synthetic method looks easier and more natural, but we should not forget that in first searching for a proof we must inevitably make use of analysis.

Thus, analysis and synthesis are two inseparably connected stages of one and the same process—the construction of a proof of a given theorem. Analysis is the method of searching for a proof; synthesis is the method of presenting the proof.

Of course, in searching for a proof, we do not always at once find the right path. It sometimes happens that we have to abandon a first approach and proceed along some other line.

Let us consider an example. Suppose that we are to prove the proposition, "If two medians of a triangle are equal, then the triangle is isosceles." Given $\triangle ABC$ in which the medians AM and BN are equal, we might first think of proving that triangles ABM and ABN are congruent. However, we see at once that we have insufficient data to prove this. We know only that $AM = BN$ and that side AB is common to the two triangles. Neither the equality of any angles nor the equality of the third side is given. Therefore, we must give up this approach. Nor can we (for lack of information) prove the congruence of triangles ACM and BCN. We, therefore, must seek a new path. Let P be the point of intersection of the medians, and consider triangles ANP and BMP. Since the two medians are equal and since, according to an earlier theorem, point P divides each median in the ratio 1:2, we find that $PN = PM$ and $PA = PB$. Moreover, $\angle APN = \angle BPM$ (vertical angles). Therefore, $\triangle ANP$ is congruent to $\triangle BMP$ and, hence, $AN = BM$. But since these segments are halves of the corresponding sides, we obtain $AC = BC$.

A facility in using the method of analysis successfully for the discovery of proofs can be developed through continued practice.

24. DIRECT AND INDIRECT PROOFS

In concluding this chapter let us consider still another distinction between methods of proof: direct and indirect proofs.

In a *direct proof* we establish the truth of the proposition to be proved by showing that it is a consequence of previously proved propositions.

In an *indirect proof,* we assume that the proposition to be proved is false, and then prove this assumption to be in contradiction either to the hypotheses or to some previously proved proposition. For this reason an indirect proof is also called a *proof by contradiction* or a *proof by reductio ad absurdum.*

For the most part the proofs given above were direct proofs. We shall now give some examples of indirect proofs.

As our first example we choose the proof of the theorem stating that two triangles are congruent if their corresponding sides are equal. The standard textbook states that it is inconvenient to prove this theorem by superposition, since we know nothing about the equality of the angles. However, by using an indirect proof, this theorem can indeed be proved by superposition.

Let ABC and $A'B'C'$ be the given triangles (Fig. 22), in which $BC = B'C'$, $CA = C'A'$, and $AB = A'B'$. For our proof let us place

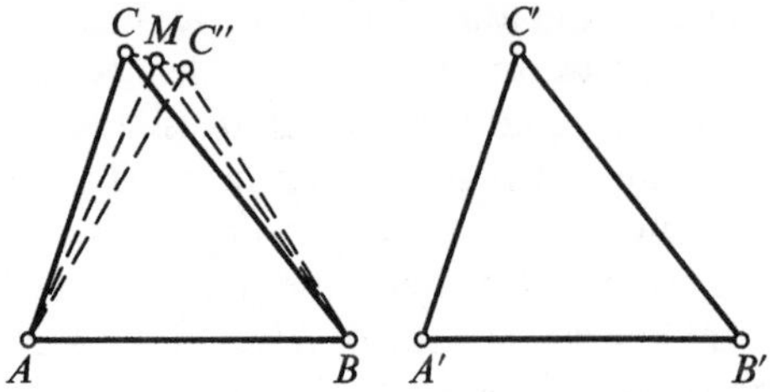

Fig. 22

$\triangle A'B'C'$ on $\triangle ABC$ in such a way that side $A'B'$ will coincide with AB. Since we know nothing about the equality of the angles, we are unable to assert that point C' will fall on point C. Let us assume, therefore, that it will come to lie at C''. We connect points C and C''. Then ACC'' is isosceles ($AC'' = AC$, by assumption); likewise, $\triangle BCC''$ is isosceles ($BC'' = BC$). The altitude AM of the isosceles triangle ACC'' will pass through the mid-point M of side CC'', since in an isosceles triangle the altitude coincides with the median. For the same reason the altitude BM of the isosceles triangle BCC'' will pass through the mid-point M of side CC''. This would mean that at point M two perpendiculars AM and BM to the straight lines CC'' have been constructed. These two perpendiculars cannot coincide, for this would mean that points A, B, and M lie in a straight line, which is impossible because of the fact that points C and C'' (and, therefore, the whole segment including point M) lie on one and the same side of the straight line AB.

Thus, on the basis of the assumption that point C' does not coincide with C, we reach the conclusion that through one and the same point M it is possible to construct two different perpendiculars to the straight line CC''. But this contradicts a previously established property of perpendiculars. Therefore, by placing $\triangle A'B'C'$ on $\triangle ABC$, point C' must coincide with point C. Hence, the two triangles are congruent.

As a second example we take the proof of the previously stated theorem (section 19) that if two angle bisectors of a triangle are equal, then the triangle is isosceles. Let there be a triangle ABC with its angle bisectors AM and BN (Fig. 23). The theorem can then be stated in the form:

Given: In $\triangle ABC$, $\angle CAM = \angle BAM$, $\angle CBN = \angle ABN$, and $AM = BN$.

To prove: $AC = BC$.

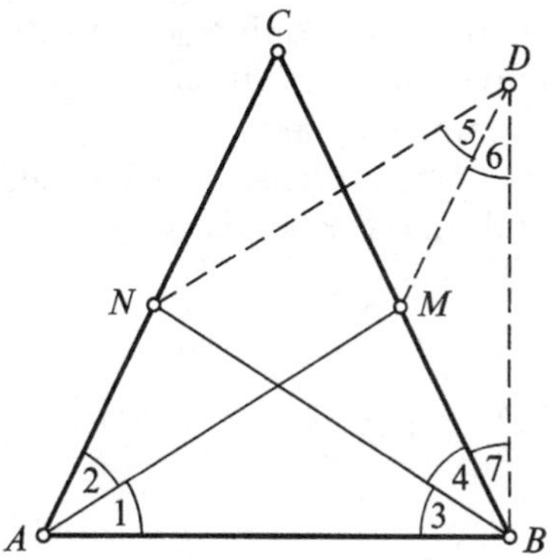

Fig. 23

We shall give an indirect proof. Accordingly, we assume that the triangle is not isosceles, that, for instance, $AC > BC$. If this is so, then $\angle ABC > \angle CAB$. Designating the angles as shown in Fig. 23, we may write $\angle 3 > \angle 1$. Now let us compare $\triangle ABM$ and $\triangle ABN$. They have side AB in common, $AM = BN$ (given) and $\angle 3 > \angle 1$. Consequently, since for any two triangles with two pairs of corresponding sides equal and the included angles unequal, the larger side lies opposite the larger included angle, we would obtain $AN > BM$.

Now, through the point N, let us draw the segment ND equal and parallel to AM. Then the quadrilateral $AMDN$ is a parallelogram. Hence, $MD = AN$ and $\angle 5 = \angle 2$. Connecting B with D, we obtain the isosceles $\triangle BDN$ ($ND = AM = BN$). On the other hand, $MD = AN$ and $AN > BM$; hence, $MD > BM$ and $\angle 7 > \angle 6$. At the same time $\angle 4 > \angle 5$, since $\angle 5 = \angle 2 = \angle 1$, and $\angle 4 = \angle 3$, while $\angle 3 > \angle 1$. If next we add the two inequalities $\angle 7 > \angle 6$ and $\angle 4 > \angle 5$, we obtain $\angle 4 + \angle 7 > \angle 5 + \angle 6$, or $\angle DBN > \angle BDN$. We have thus been led to the conclusion that the base angles of the *isosceles* $\triangle BDN$ are unequal. Since this false conclusion followed from our assumption $AC > BC$, that assumption must have been false. Quite analogously we see that a contradiction arises if we assume that $BC > AC$. Hence, $AC = BC$.

These two examples may suffice to characterize the nature of indirect proofs. This type of proof is often used when, in the search for arguments, it appears that a direct proof will be difficult, or even impossible, to find. In such cases we assume that the proposition contrary to the one to be proved is true and then try to find a chain of reasoning leading to a conclusion contradicting some previously established proposition. Thus, in the first of our two examples we arrived at the conclusion that through a single point two perpendiculars to a straight line could be drawn; in the second, that the base angles of an isosceles triangle are not equal.

4. What Propositions in Geometry Are Accepted without Proof?

25. BASES FOR SELECTION OF AXIOMS

What propositions in geometry are accepted without proof? At first glance this question appears very simple. You will say that we accept as axioms those propositions whose truth has long been verified and about which there is no doubt.[1] However, when we try to select such propositions in actual practice, we find that it is not so simple as it seems. A very large number of geometric propositions have been subjected to practical verification so often that scarcely anyone doubts their truth. But this does not mean that we should regard all such propositions as axioms. For example, we know that through two points only one straight line can be drawn; that at a given point one and only one perpendicular can be drawn to a straight line; that the sum of two sides of a triangle is larger than the third side; that two segments which are equal to a third are equal to each other; that the distance between two parallel straight lines is everywhere the same; etc. The number of such obviously true propositions could be multiplied many times over. Why not accept all such propositions as axioms? Would not the exposition of geometry be greatly simplified in that way, with many proofs becoming superfluous?

In fact, the development of geometry has not moved along these lines. On the contrary, geometers have sought to reduce the list of axioms to the smallest number possible and to obtain the entire remaining content of geometry by deductive reasoning from this small number of fundamental truths.

Why did they choose this apparently more difficult and complicated way of constructing a system of geometric knowledge? This effort was made for a number of reasons. First of all, when the number of axioms is reduced, the significance of each individual axiom increases, for these axioms contain within themselves,

[1] *Editor's note.* Compare, however, pp. 19–21 in the book by Wilder mentioned in the note on p. 6.

so to speak, the whole future geometry to be deduced from them. Therefore, the fewer the axioms, the more far-reaching and profound and important are the properties of spatial forms revealed by each individual axiom.

Another reason for striving to limit the number of axioms is that the smaller the number of axioms, the easier it is to examine the validity of each axiom and the validity of all the axioms combined. Thus, we are faced with the problem of selecting the *smallest possible number* of the most basic, general, and important propositions of geometry to be accepted as axioms. How are we to make this selection?

26. PROPERTIES OF A SYSTEM OF AXIOMS

In the first place, we have to keep in mind that we cannot select the axioms randomly one after the other; instead, they will have certain relations to one another. Thus, in geometry we have not single, isolated axioms, but rather *a whole system of axioms,* since only a system as a whole can correctly represent the existing properties and interrelationships of the fundamental spatial forms of the material world.

Furthermore, in choosing a system of axioms, we must take care that no two propositions should be included which contradict each other, since such propositions cannot both be true. It would be impossible, for example, to admit to the system the two axioms: "Through a given point it is possible to draw one and only one parallel to a given straight line" and "Through a given point no line can be drawn which is parallel to a given straight line." Moreover, not only must the axioms as such be compatible with one another, but among the propositions deduced from the axioms there must not be any two which contradict each other. This fundamental requirement is called the *condition of consistency.*

In addition to this we must also take care that no proposition is included in our system of axioms which can be deduced from other axioms. This requirement is obvious in view of the fact that we wish to make our system a minimal one, that is, containing the smallest possible number of unproved propositions. If a given proposition can be proved from other axioms, then it is not an axiom but a theorem, and there is no justification for including it in the system of axioms. The requirement that no axiom be deducible from other axioms is called the *condition of independence.*

However, in our endeavor to make the system of axioms minimal, we must not go to extremes and omit from it any propositions which are needed in deducing any of the theorems of geometry. This leads us to the third condition which a system of axioms must satisfy—*the condition of completeness.* This condition may be formulated more precisely thus: If the system of axioms is *incomplete* then it is always possible to construct a new proposition (a proposition employing, of course, the same fundamental concepts as the axioms) which is not deducible from the axioms and which does not contradict them either. If, however, the system of axioms is *complete,* then any new proposition, added to the system of geometry and using the same concepts as those with which the axioms deal, either will be a consequence of these axioms or will contradict them.

27. ANALOGY FROM ALGEBRA

In order to elucidate further the significance of these three conditions of completeness, independence, and consistency of a system of axioms, we give the following simple example which, although it is not an expression of geometric relationships, provides a fairly good analogy to them.

Let us examine a system of first-degree equations in three unknowns. We shall consider each of the unknowns of the system as a "concept," subject to definition, and each equation as a sort of "axiom" by means of which the relations between these "concepts" can be determined. Thus, let us take the system

$$\begin{aligned} 2x - y - 2z &= 3, \\ x + y + 4z &= 6. \end{aligned}$$

Is it possible to determine the unknowns x, y, and z from this system? No, since in this case the number of equations is less than the number of unknowns. The system does not satisfy the *condition of completeness.*

Now let us try to improve this system by supplementing it with one more equation:

$$\begin{aligned} 2x - \ y - \ 2z &= 3, \\ x + \ y + \ 4z &= 6, \\ 3x + 3y + 12z &= 18. \end{aligned}$$

By carefully examining the new system, we find that the introduction of the new equation has not changed the situation, since the third equation follows directly from the second (the third equation is just the second equation multiplied by three) and does not supply any new relationships. The system violates the *condition of independence*.

Now let us change the third equation and examine the following system:

$$\begin{aligned} 2x - \ y - \ 2z &= 3, \\ x + \ y + \ 4z &= 6, \\ 3x + 3y + 12z &= 15. \end{aligned}$$

But again this system is useless for the determination of the unknowns. For dividing both sides of the last equation by 3, we get the equation

$$x + y + 4z = 5,$$

while the second equation gives us

$$x + y + 4z = 6.$$

Which of these equations are we to believe? It is clear that this is a case of an *inconsistent* system, from which it is likewise impossible to determine the unknowns.

If, finally, we examine the system

$$\begin{aligned} 2x - y - 2z &= 3, \\ x + y + 4z &= 6, \\ 2x + y + 5z &= 8, \end{aligned}$$

it is easy to see that the system does have a single solution ($x = 5$, $y = 13$, $z = -3$); it is *consistent, independent,* and *complete.*

If we add to this system a fourth equation in x, y, and z, it will either be a combination of the three given equations or else it will contradict them.

From all these considerations we see that the selection of the axioms which are to be used as the foundation of geometry is far from arbitrary and is subject to very rigid requirements. The work of determining an acceptable system of axioms for geometry was begun near the end of the last century, and although scholars have made much progress in this direction, it cannot be considered as completed even at the present time. In particular, in subjecting an existing system of axioms to systematic re-examination, mathematicians at times discover that the system contains superfluous, that is, "dependent," axioms, which can be deduced from simpler or more general ones. All these investigations are of great interest to the mathematician because their purpose is to ascertain which most general, basic, and important properties of spatial forms determine the entire content of geometry.

In order to give the reader some idea of the system of axioms of contemporary geometry, let us turn first to the exposition of geometry in school and see upon what axioms it is constructed and which axioms are lacking. We shall limit ourselves to the axioms of plane geometry.

28. AXIOMS OF CONNECTION

The school course of geometry begins by introducing the basic concepts of geometry: solids, surfaces, lines, points. Next the straight line is singled out from among all other lines, and the plane from all other surfaces.

The first axioms of the school course lay down the relations between points, straight lines, and planes. These axioms belong to the group of *axioms of connection*—the first group in the complete system of geometric axioms. The axioms of this group state how the fundamental geometric entities are "connected" with each other—by how many points a straight line and a plane are determined, under what conditions a straight line lies in a plane, etc. Of these axioms of connection only two are mentioned in the school course:

1) *Through two points one and only one straight line can be drawn.*

2) *If two points of a straight line lie in a plane, then the entire straight line lies in that plane.*

But we also constantly make use, consciously or unconsciously, of other axioms of connection, of which the following are needed as a basis for plane geometry:

3) *Each straight line contains at least two points.* This axiom, as we see, lays down a minimal requirement. Afterwards, on the basis of the axioms of order, the existence of an infinite number of points on a straight line can be proved by means of this axiom.

4) *There exist in a plane at least three points which do not lie on one and the same straight line.* This axiom also contains a minimal requirement, on the basis of which we can later prove the existence of an infinite number of points in a plane.

29. AXIOMS OF ORDER

Let us now turn to the second group of axioms, which are sometimes absent from the school course, although one has to make use of them at every step. The axioms of the second group are called *axioms of order*. These axioms describe the laws which govern the relative position of points on a straight line and the mutual position of points and straight lines in a plane.

We often use these axioms, although usually not in an explicit form. If, for example, we have to extend a line segment, we do this knowing that a segment can always be extended in either direction. If we join two points that lie on different sides of a straight line, we are confident that the resulting line segment will intersect the straight line. We relied on this, for example, in the proof of the theorem on the congruency of triangles with two sides and an angle lying opposite one of these sides correspondingly equal (see Fig. 12). Or also, we are sure that the bisector of an interior angle of a triangle will, under all circumstances, intersect the opposite side. Unquestionably, in all these cases we deal with what seem to be very obvious facts. But they concern the existence of certain basic properties of geometric figures, of which we make constant use in our deductions and which, therefore, must be set forth as axioms.

The axioms which refer to the order of points on a straight line are expressed in terms of the basic concepts "to precede" and "to follow." They are formulated as follows:

1) *Of any two points A and B on a straight line, either one may be considered as "preceding" the other; if A "precedes" B, then B "follows" A.*

2) *If A, B, C are points on a straight line, and if A precedes B, and B precedes C, then A precedes C.*

These two axioms alone already describe characteristic properties of a straight line which do not pertain to all lines (curved, in general). For example, moving clockwise around a circle (Fig. 24) and marking successively on it the points *A*, *B*, and *C*, we find that point *A* precedes point *B*, point *B* precedes point *C*, but point *C* again precedes point *A*.

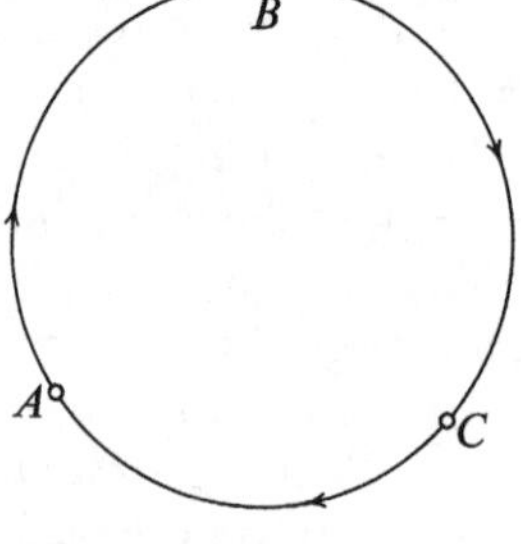

Fig. 24

Of the three points *A, B,* and *C* on the straight line mentioned above, we say that *B* lies *between A* and *C* (Fig. 25).

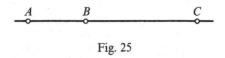

Fig. 25

3) *Between any two points on a straight line, there exists another point on the same straight line.*

Using this axiom repeatedly, first for two points on a straight line (which do exist by virtue of the second axiom of connection), and next for each of the points obtained between them, and so on, we deduce that between any two points on a straight line there exists an infinite number of points on this same straight line.

The part of a straight line to which two of its points and all points between belong is called a *segment.*

4) *For every point on a straight line there exist both a preceding point and a following point.* From this axiom it follows that it is possible to produce a segment of a straight line in either direction. From this it also follows that on a straight line there is no point which precedes all its other points, nor which follows all its other points; that is, *a straight line has no ends.*

The part of a straight line which contains a given point and all those which precede it, or a given point and all those which follow it, is called a *ray* or *half-line.*

The relative position of points and straight lines in a plane is determined by the following axiom, which is called *Pasch's axiom,* after the German mathematician who first formulated it:

5) *If three points are given which do not lie in a straight line, then a straight line in the same plane, which does not pass through any of these three points but which intersects one of the segments determined by them, intersects one and only one of the other two segments determined by the three points* (Fig. 26).

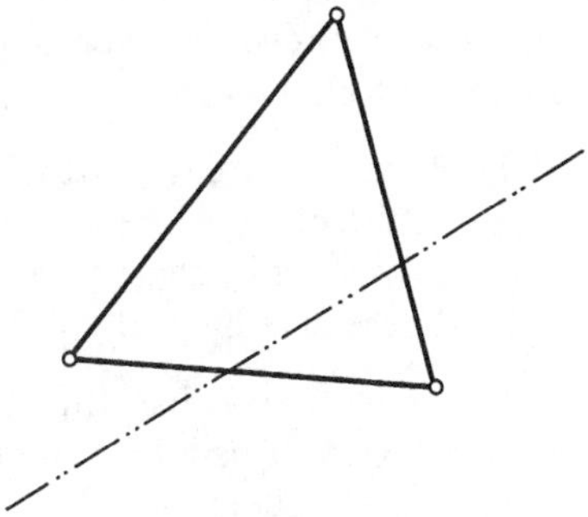

Fig. 26

With the aid of this axiom we can prove the theorem that a straight line divides a plane into two half-planes. We shall give this proof here as an example of a rigorous proof relying only on axioms and previously proved propositions. Let us formulate the theorem as follows:

Any straight line which lies in a plane divides all the points of the plane which do not belong to it into two sets such that two points of one and the same set determine a segment which does not intersect the straight line, while two points of different sets determine a segment which does intersect the straight line.

For the sake of brevity, we shall make use of certain symbols. The symbol $\in$ means *belongs to:* $A \in a$ means "point A belongs to straight line a." The symbol $\times$ means *intersects:* $AB \times a$ means "segment AB intersects straight line a." A dash above any relationship denotes *negation;* for example, $\overline{A \in a}$ means "point A does *not* belong to the straight line a." Now, we proceed with the proof of the theorem.

First, let us note that *if three points lie in a straight line,* then a proposition holds true for them which is analogous to Pasch's axiom: *A straight line which intersects one of the three segments determined by these points, and does not pass through any of the three points, intersects one and only one of the other two segments.*

This proposition is readily proved by using the axioms of order for points on a straight line. Indeed, if points A, B, and C lie in a straight line and point B lies between A and C, then all the points of segments AB and BC belong to segment AC, and any point on segment AC (except B) belongs to either AB or BC but not to both. Therefore, a straight line intersecting AB or BC (but not at A, B, or C) certainly intersects AC also, and a straight line intersecting AC (but not at A, B, or C) intersects either AB or BC but not both.

Now let there be given a straight line l in a plane. We are to prove the following:

1) By means of the straight line l it is possible to divide the points of the plane which do not belong to l into sets.

2) There can be two and only two such sets.

3) These sets have the properties pointed out in the theorem.

In order to establish this, let us take point A (Fig. 27) not on the line l, and let us adopt the following terminology:

a) point A belongs to the *first* set (designated as K_1);

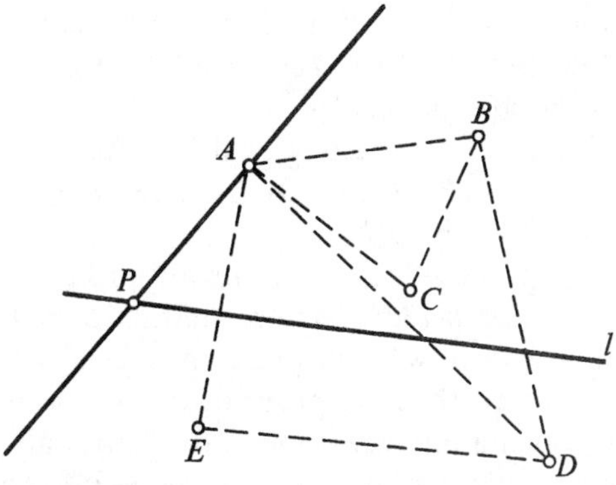

Fig. 27

b) a point which does not belong to l belongs to the *first* set if, together with A, it determines a segment which does not intersect l;

c) a point which does not belong to l belongs to the *second* set (designated as K_2) if, together with A, it determines a segment which intersects l.

We first prove that points of each of these sets *exist*. In order to do so, let us take point P on line l and draw the straight line PA. The half-line with vertex P, which contains point A, contains only points of the first set, since the point of intersection P with l lies outside of the segments which are determined by point A and the remaining points of the half-line. The oppositely directed half-line from P contains only points of the second set, since the point of intersection P with l lies inside all segments determined by point A and by points of this half-line. Joining A with any point of l, we get an infinite number of straight lines containing points of the first and second sets.

There can be *only two* sets, since about any segment that connects A with a point outside l, we can make *only two* assertions—either the segment intersects l or it does not intersect l; there is no third alternative.

Finally, we show that the sets K_1 and K_2 satisfy the conditions of the theorem. Let us consider the following cases:

1) Two points B and C both belong to the first set; that is, $B \in K_1$ and $C \in K_1$. Since $B \in K_1$, $\overline{AB \times l}$ and $C \in K_1$, $\overline{AC \times l}$, we have on the basis of Pasch's axiom, $\overline{BC \times l}$.

2) Two points D and E both belong to the second class; that is, $D \in K_2$ and $E \in K_2$. Since $D \in K_2$, $\overline{AD \times l}$ and $E \in K_2$, $AE \times l$, we have, on the basis of Pasch's axiom, $\overline{DE \times l}$.

3) Two points B and D belong to different classes, $B \in K_1$ and $D \in K_2$. Since $B \in K_1$, $\overline{AB \times l}$ and $D \in K_2$, $AD \times l$, we have, on the basis of Pasch's axiom, $BD \times l$.

The part of a plane which contains all points of one and the same set is called a *half-plane*. Thus, the theorem that a straight line divides a plane into two half-planes has been proved.

It should be noted that the proof of this theorem can be carried out without using any diagram. A diagram only helps to follow the reasoning more easily. This observation applies to any truly rigorous proof.

30. AXIOMS OF CONGRUENCE

The next, or third, group of geometric axioms refers to the concept of *congruence*. In the school course in geometry the congruence of figures in a plane is determined by the superposition of one figure on the other. Most ordinary geometry textbooks make the following statement in regard to this subject: "Geometric figures may be moved about in space without undergoing any changes. Two geometric figures are called congruent if, by moving one of them in space it may be made to coincide with the second figure in such a way that both figures coincide in all their parts."

At first glance this definition of congruence seems perfectly clear. But if it is carefully analyzed it shows circular reasoning. Indeed, to determine the congruence of figures we must make them coincide; and to make them coincide we must move one figure in space, claiming that during the process of being moved it remains *unchanged.* But what does it mean to "remain unchanged"? It means that the figure always remains congruent to its original shape. Thus, we define the concept "congruence" by means of moving an "unchanging figure," and define the concept of an "unchanging figure" by means of the concept of "congruence." Therefore, it appears to be much better to define the congruence of figures by means of a group of axioms concerning the equality of segments, angles, and triangles.

The axioms concerning the equality of segments are as follows:

1) *On a given straight line in a given direction from a given point, it is possible to lay off one and only one segment equal to a given one.*

2) *Any segment is equal to itself. If a first segment is equal to a second, then the second is equal to the first. Two segments, equal to the same third segment, are equal to each other.*

3) *If A, B, and C lie in a straight line, and A', B', and C' also lie in a straight line, and if $AB = A'B'$ and $BC = B'C'$, then* $AC = A'C'$. In other words, if to equal segments we add equal ones, then the sums will also be equal.

Completely analogous axioms exist for angles.

4) *On a given half-line and in one of the two half-planes determined by the line, it is possible to construct one and only one angle equal to a given one.*

5) *Every angle is equal to itself. If a first angle is equal to a second, then the second is equal to the first. If two angles are equal to the same third angle, then they are equal to each other.*

6) *If a, b, and c are half-lines with a common vertex, and a', b', and c' are other half-lines with a common vertex, and if $\angle ab = \angle a'b'$ and $\angle bc = \angle b'c'$, then $\angle ac = \angle a'c'$.* In other words, if to equal angles we add equal angles, then the sums will also be equal.

Finally, to establish the congruence of triangles still another axiom is introduced into the third group.

7) *If two sides and the angle included between them in one triangle are respectively equal to two sides and the angle included between them in another triangle, then, in these triangles, the other angles are correspondingly equal.* For example, if we have $\triangle ABC$ and $\triangle A'B'C'$, and if $AB = A'B'$, $AC = A'C'$, and $\angle A = \angle A'$, then $\angle B = \angle B'$ and $\angle C = \angle C'$.

The fundamental theorem of congruence of triangles can then be proved on the basis of these seven axioms, as can all the theorems on the congruence of figures which are based on them. In this method, therefore, there is no need for using the idea of superposition.

Let us see how, for example, one of the well-known theorems on congruence of triangles can now be proved. Let $\triangle ABC$ and $\triangle A'B'C'$ (Fig. 28) be given, in which $AB = A'B'$, $AC = A'C'$, and $\angle A = \angle A'$. We are to prove that all the remaining elements of the triangles are also respectively equal. From axiom 7 we obtain at once $\angle B = \angle B'$ and $\angle C = \angle C'$. There remains to be proved

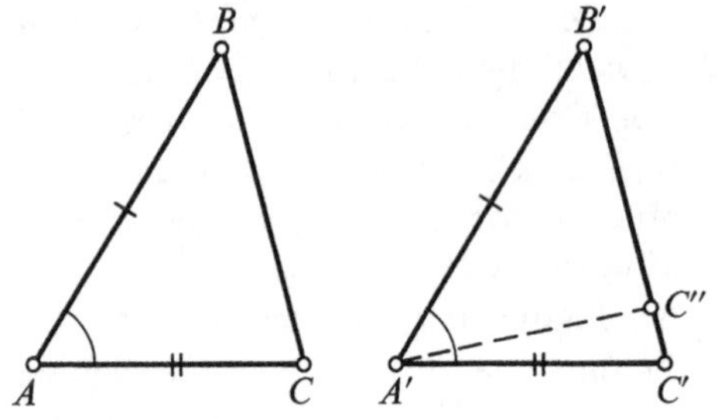

Fig. 28

$BC = B'C'$. Let us assume that $BC \neq B'C'$. Then on side $B'C'$, from point B', we can lay off $B'C'' = BC$. Now we examine $\triangle ABC$ and $\triangle A'B'C''$. In these triangles $AB = A'B'$, $BC = B'C''$ and $\angle B = \angle B'$. Then, according to axiom 7, $\angle B'A'C'' = \angle A$. But two angles which are equal to the same third angle are equal to each other; therefore, $\angle B'A'C'' = \angle B'A'C'$. This would mean that on the half-line $A'B'$, in one and the same half-plane, two different angles have been constructed equal to the same angle A. This contradicts axiom 4. Thus, having disproved the assumption that $BC \neq B'C'$, we have proved $BC = B'C'$.

The other theorems of the congruence of figures are proved in similar fashion.

31. AXIOMS OF CONTINUITY

In the course of the further development of elementary geometry the need arises for introducing one more group of axioms, the *axioms of continuity*. Problems connected with the intersection of a straight line with a circle and the intersection of circles with each other involve this group of axioms. The fact that all geometric constructions with compass and ruler are based on these very problems suggests the extreme importance of the axioms of continuity. Indeed, the entire theory of the measurement of geometric magnitudes is based on the axioms of continuity.

The following are two axioms from this group:

1) THE AXIOM OF ARCHIMEDES. *If two segments are given, of which the first is larger than the second, then, by laying off the smaller segment a sufficient number of times on the larger one, we can always obtain a sum which exceeds the larger segment.* Or also, if $\bar{a}$ and $\bar{b}$ are two segments such that $\bar{a} > \bar{b}$, then there exists an integer n such that $n\bar{b} > \bar{a}$.

The method of finding the common measure of two segments by repeatedly laying off remainders, which we discussed above, is based on the axiom of Archimedes. For in this method the smaller segment is repeatedly laid off on the larger, and the axiom of Archimedes assures us that in this way the sum of the smaller segments will ultimately exceed the larger segment. We conclude directly from the axiom of Archimedes that if segment $\bar{a}$ is larger than segment $\bar{b}$ then there always exists an integer n such that $\frac{\bar{a}}{n} < \bar{b}$.

A second axiom of continuity is called CANTOR'S AXIOM or the *axiom of closed nested intervals:*

2) *If a sequence of closed intervals*[1] *is such that each one is located within the preceding one, and if in this sequence it is always possible to find an interval whose length is less than that of any arbitrary preassigned interval, then there exists a unique point lying within each of these intervals.*

As an application of Cantor's axiom, let us consider the following example: We take a segment A_0B_0 (Fig. 29). Let its mid-point

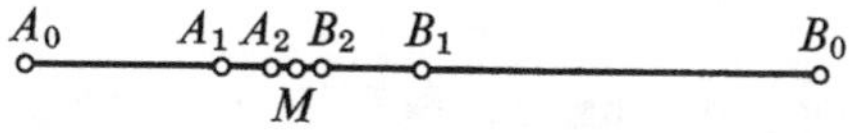

Fig. 29

be B_1; next we take the mid-point of the segment A_0B_1, which we call A_1; next we take the mid-point of A_1B_1, which we call B_2; next the mid-point of the segment A_1B_2, which we call A_2; next the mid-point of A_2B_2, which we call B_3, and again the mid-point of A_2B_3, which we call A_3; then the mid-point of A_3B_3; and so on. The segments A_0B_0, A_1B_1, A_2B_2, A_3B_3, . . . , etc., constitute a sys-

[1] A *closed interval* is an interval including its end points.

tem of "closed nested intervals." Each segment is located within the preceding one and is equal to $\frac{1}{4}$ of the preceding one. Thus, the length of segment A_1B_1 is equal to $\frac{1}{4}A_0B_0$, the length of A_2B_2 $=\frac{1}{16}A_0B_0$, $A_3B_3 = \frac{1}{64}A_0B_0, \ldots$, and, in general, $A_nB_n = \frac{A_0B_0}{4^n}$.

From the axiom of Archimedes it follows that the length $\frac{A_0B_0}{4^n}$ when n is sufficiently large may be made less than the length of any given segment. Thus, all conditions of Cantor's axiom are fulfilled, and there exists a unique point lying within each of the segments. As we shall prove below, this point has the distance $\frac{1}{3}A_0B_0$ from A_0; or, denoting the point by M, we shall prove that $A_0M = \frac{1}{3}A_0B_0$. If we take the length of segment A_0B_0 as 1, then the distances d_i (where $i = 1, 2, 3, \ldots, n$) of the points $A_1, A_2, A_3, \ldots, A_n$ from A_0 are, respectively,

$$d_1 = \frac{1}{4},$$

$$d_2 = \frac{1}{4} + \frac{1}{4^2} = \frac{5}{16},$$

$$d_3 = \frac{1}{4} + \frac{1}{4^2} + \frac{1}{4^3} = \frac{21}{64}, \ldots,$$

$$d_n = \frac{1 + 4 + 4^2 + \cdots + 4^{n-1}}{4^n}.$$

We shall now show that each of these fractions is less than $\frac{1}{3}$. For, if the denominator of each of them is diminished by one, this increased fraction becomes exactly equal to $\frac{1}{3}$:[1]

$$d_n < \frac{1 + 4 + 4^2 + \cdots + 4^{n-1}}{4^n - 1}$$

$$= \frac{1 + 4 + 4^2 + \cdots + 4^{n-1}}{(4-1)(1 + 4 + 4^2 + \cdots + 4^{n-1})} = \frac{1}{3}.$$

[1] Here we make use of the formula
$$a^n - b^n = (a - b)(a^{n-1} + a^{n-2}b + a^{n-3}b^2 + \cdots + ab^{n-2} + b^{n-1}).$$

On the other hand, the distances D_i of the points B_1, B_2, B_3, $\ldots$, B_n from A_0 are, respectively,

$$D_1 = \frac{1}{2},$$

$$D_2 = \frac{1}{2} - \frac{1}{8} = \frac{3}{8},$$

$$D_3 = \frac{1}{2} - \frac{1}{8} - \frac{1}{32} = \frac{11}{32}, \ldots,$$

$$D_n = \frac{1}{2} - \frac{1}{8} - \frac{1}{32} - \cdots - \frac{1}{2^{2n+1}}.$$

We next write the distance D_n in the form

$$D_n = \frac{1}{2} - \left(\frac{1}{4} - \frac{1}{8}\right) - \left(\frac{1}{16} - \frac{1}{32}\right) - \cdots - \left(\frac{1}{2^{2n}} - \frac{1}{2^{2n+1}}\right)$$
$$= \frac{1}{2} - \frac{1}{4} + \frac{1}{8} - \frac{1}{16} + \frac{1}{32} - \cdots - \frac{1}{2^{2n}} + \frac{1}{2^{2n+1}}.$$

If we add these terms over a common denominator, we obtain

$$D_n = \frac{2^{2n} - 2^{2n-1} + 2^{2n-2} - \cdots - 2^3 + 2^2 - 2 + 1}{2^{2n+1}}.$$

From this we can show that each of the distances $D_1, D_2, \ldots, D_n$ is greater than $\frac{1}{3}$. For, by increasing the denominator of the fraction by one and thereby decreasing the value of the fraction, we obtain[1]

$$D_n > \frac{2^{2n} - 2^{2n-1} + 2^{2n-2} - \cdots - 2^3 + 2^2 - 2 + 1}{2^{2n+1} + 1}$$
$$= \frac{2^{2n} - 2^{2n-1} + 2^{2n-2} - \cdots - 2^3 + 2^2 - 2 + 1}{(2 + 1)(2^{2n} - 2^{2n-1} + 2^{2n-2} - \cdots - 2^3 + 2^2 - 2 + 1)} = \frac{1}{3}.$$

Thus, all distances $D_1, D_2, D_3, \ldots, D_n, \ldots$ of the points B_1, B_2, B_3, $\ldots$, $B_n, \ldots$ from A_0 are larger than $\frac{1}{3}$. From this it follows that a point M at the distance $\frac{1}{3}$ from A_0 lies within each of the intervals $A_1B_1, A_2B_2, \ldots, A_nB_n, \ldots$. Hence, it is the unique point determined by the sequence of these segments.

[1] Here we make use of the formula

$$a^{2n+1} + b^{2n+1} = (a + b)(a^{2n} - a^{2n-1}b + a^{2n-2}b^2 - \cdots - ab^{2n-1} + b^{2n}).$$

32. THEOREMS BASED ON THE AXIOMS OF CONTINUITY

We shall now proceed to the proof of the basic theorem regarding the intersection of a straight line with a circle. We recall that a circle is determined by its center and radius. The points of the plane whose distance from the center is less than the radius are called *interior* points of the circle; the points whose distance from the center is greater than the radius are called points *exterior* to the circle. The basic theorem is formulated as follows:

A segment that joins an interior point of a circle with an exterior one has one and only one point in common with the circle.

Suppose we are given a circle with center O and radius r, with an interior point A ($OA < r$) and an exterior point B ($OB > r$) (Fig. 30). We begin by proving that if on the segment AB there exists a point M, whose distance from O is equal to the radius, then there can be only one such point. Indeed, if such a point M exists, then a second point M' also exists, symmetrical to M with respect to the perpendicular dropped from O to the straight line AB ($M'O = MO = r$). In accordance with the property of oblique lines from a point to the straight line AB, all interior points of the segment $M'M$ will also be interior points of the circle, while all points exterior to the segment $M'M$ will also be exterior points of the circle. Therefore, point A must always lie between points M' and M, and hence segment AB can contain only the one point M. That is, segment AB cannot intersect the circle in more than one point.

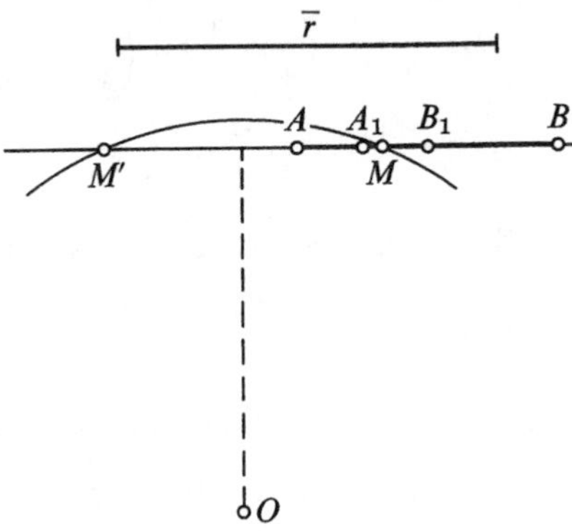

Fig. 30

We next prove that AB must intersect the circle in one point. For this purpose we bisect segment AB and compare the distance from the mid-point to the center O with the radius of the circle. If this distance is equal to the radius, then the theorem is proved. If this distance is less than the radius, then the mid-point is an interior point; we call it A_1. If this distance is greater than the radius, then the point is an exterior point; we call it B_1.

Next, we take the mid-point of the segment A_1B (or AB_1), with respect to which there are again three possibilities—its distance from the center is equal to the radius, in which case the theorem is proved; or it is less than the radius, in which case we designate the point by the letter A with the corresponding subscript; or it is greater than the radius, in which case we designate it with the letter B with the corresponding subscript. Continuing this process without limit, we always find that either the distance of any such mid-point from the center O is equal to the radius, in which case the theorem is proved, or that all points designated by $A_1, A_2, \ldots, A_n, \ldots$ are interior points and those designated by $B_1, B_2, \ldots, B_n, \ldots$ are exterior points. In this latter case, however, we have a sequence of segments which satisfy the conditions of Cantor's axiom, since each successive segment lies within the preceding one, and the length of each succeeding segment is half that of the preceding one. This means that there exists a unique point lying within all these segments. Since it lies between all interior and all exterior points of the segment, it can be neither an interior nor an exterior point. Hence, it is a point on the circle.

From this theorem we can next prove that *if the distance of a straight line from the center of a circle is less than the radius, then the straight line has two and only two points in common with the circle.* Let O be the center and r the radius of a circle (Fig. 31). Since the distance OP from the center to the straight line l is less than the radius, P is an interior point. Now from point P let us lay off on the straight line l a segment $PQ = r$.

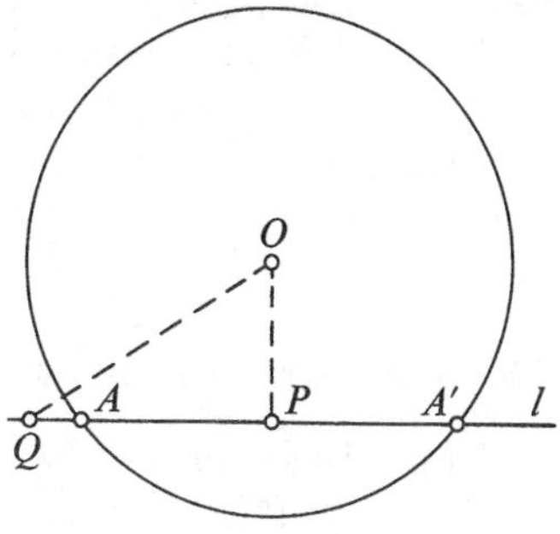

Fig. 31

Since in the right triangle OPQ the hypotenuse OQ is greater than the leg $PQ = r$, we have $OQ > r$, so that Q is an exterior point. According to the theorem just proved, the segment PQ has a single point A in common with the circle. A second common point A' on the line l is the one symmetric to A with respect to the perpendicular OP. Since all interior points of the segment AA' are also interior points of the circle, and all points exterior to the segment are also exterior to the circle, the straight line l has no other points in common with the circle.

Propositions analogous to the axioms of Archimedes and Cantor referring to arcs of a circle appear now as theorems. Thus, we can prove that

1) *By laying off a given arc a sufficient number of times, we can obtain an arc greater than any previously given arc.*

2) *If there exists a sequence of arcs, in which each successive arc lies within the preceding one, and if in this sequence we can always find an arc smaller than any given arc, then there exists a point lying within all these arcs.*

By using these propositions it is easy to prove the basic theorem on the intersection of circles:

If A is an interior point and B an exterior point of a given circle, then the arc of any other circle connecting A and B has one and only one common point with the given circle.

The proof of this theorem is entirely analogous to the proof of the theorem on the intersection of a circle with a line segment.

33. AXIOM OF PARALLELISM

The fifth and last group of axioms of geometry is connected with the concept of *parallelism* and contains only one axiom.

Through a point outside a given straight line there exists one and only one straight line parallel to the given line.

Propositions based on this axiom are well known, and we shall not dwell on them.

34. REDUCTION OF THE NUMBER OF AXIOMS

This discussion of the axioms will have given a sufficiently clear idea of this system of unproved propositions which constitutes the foundation of geometry. It should be noted, however, that in trying to simplify our presentation as much as possible, we have not tried to make the system minimal in number. In fact, the number of these axioms could be decreased further. For example, the two axioms of Archimedes and Cantor can be replaced by a single one, the so-called axiom of Dedekind. Also, the assertions of some of the axioms could be reduced somewhat. For example, in Pasch's axiom it is not necessary to postulate that a straight line which intersects one of the sides of a triangle intersects one and *only* one of the other sides. It is enough to postulate that it intersects one other side of the triangle; the fact that it intersects only one of the other sides can be proved. Similarly, in the formulation of Cantor's axiom we need not postulate that a point determined by a system of closed nested intervals is unique; the uniqueness of this point can be proved. However, all this would complicate and lengthen our exposition.

35. SUMMARY

Let us then summarize what we have said in this book.

1) We have defined geometry as the science of the spatial forms of the material world.

2) We have pointed out that our primary knowledge of the properties of the spatial forms is obtained by means of induction, that is, from observations and experiments.

3) We have formulated the most basic and most general spatial properties of objects in the form of a system of fundamental propositions—axioms.

4) A correctly constructed system of axioms must satisfy the conditions of completeness, independence, and consistency.

5) Aside from the axioms, all propositions of geometry (the theorems) are derived by chains of deduction, that is, by means of deductive reasoning from axioms and previously proved theorems. Such chains of deductions are called proofs.

6) In order that a proof be correct, it must be based on correct reasoning. Success in giving correct proofs depends on 1) a precise and correct formulation of the proposition to be proved, 2) the selection of necessary and valid arguments, and 3) a strict observance of the rules of logic in the course of the proof.

Mistakes in Geometric Proofs

by Ya. S. Dubnov

PREFACE TO THE AMERICAN EDITION

THIS BOOKLET presents examples of faulty geometric proofs, some of which illustrate mistakes in reasoning that a student might make, while others are classic sophisms. Chapters 1 and 3 present these faulty proofs, and then Chapters 2 and 4 give detailed analyses of the mistakes.

Naturally, in order to read this booklet, the reader must be acquainted with plane geometry. Only an acquaintance with theorems concerning parallel and perpendicular lines and polygons is needed for Chapters 1 and 2. Chapters 3 and 4 contain more advanced material and presuppose some knowledge of the simpler properties of circles, the concept of limit, trigonometry, and some solid geometry.

It is suggested that the reader first examine the examples of incorrect proofs given in Chapters 1 and 3. He should attempt to discover the mistakes in these examples by himself before reading Chapters 2 and 4. Portions of the text appearing in fine print, as well as many of the footnotes, may be omitted on first reading; these are intended primarily for the more advanced reader.

CONTENTS

Introduction

Forty years ago in an article dealing with the teaching of geometry, a then well-known Russian mathematician and teacher, N. A. Izvolskii, reported a conversation which he had had with a school-girl acquaintance who had just been studying geometry for one year. The teacher asked her what she remembered from the geometry course. After thinking for a long time, the girl was unfortunately unable to remember anything. The question was then recast in another form: "What did you do during the year in your geometry lessons?" This brought the very quick reply, "We did proofs."

This answer reflects an idea widely held among pupils—in arithmetic problems are solved, in algebra equations are solved and formulas are derived, but in geometry theorems are proved. This conception of mathematics was correct a long time ago, but in present-day mathematical studies, theorems followed by proofs are encountered equally often whether one is dealing with numbers or diagrams. Moreover, problems are solved in all fields of mathematics, and in geometry we often solve equations. It was different 2,000 years ago when Euclidean geometry, which still forms the basis of the high school geometry course, was being created. From that time up to the present day, geometry (but not the other branches of mathematics) has been expounded in textbooks in the form of a chain of theorems (some of which are called lemmas or corollaries). These are constructed according to a plan which is so well known that we limit ourselves to a short reminder. Each theorem contains a condition ("Given . . .") and a conclusion ("Prove . . ."); in the proof we may rely only on axioms or on theorems already proved.

The part which drawings play in proofs is well known; they clarify not only the content of the theorem but also the course of the proof. Sometimes several drawings are required for a single theorem, because the proof may vary with the relative positions of the parts of a figure. For example, the proof of a theorem about an angle inscribed in a circle usually involves three possibilities: the center of the circle lies on a side of the angle, inside of the angle, or outside of it. It is important to exhaust all possible arrangements of the parts of the figure; the omission of any one arrangement to which the exhibited reasoning cannot be applied, of course, invalidates the entire proof—it may be precisely for that one arrangement that the theorem turns out to be false.

The role of the drawing should be neither exaggerated nor underestimated. To regard it as an indispensable part of the proof would be an exaggeration. Theoretically speaking, a geometric proof does not need drawings; not using them even has the advantage of removing any reliance on what is "self-evident" from the drawing, which sometimes is only apparently "self-evident" and is a source of error. In practice, however, dispensing with the drawing might lead to the same difficulties as we would experience if, for instance, we were to try to perform computations with numbers having several digits completely in our heads, or if we were to play chess without looking at the chess board; the danger of making a mistake would be increased considerably.

In speaking of the help afforded by a drawing in developing a proof, I have in mind, of course, a good drawing carefully executed. A bad drawing can be a hindrance. In this booklet the reader will encounter not only correct drawings, but also others which are somewhat distorted. This is done deliberately; we are concerned here with faulty proofs, and these sometimes result from inaccurate drawings.

In Chapters 1 and 3, a number of examples of faulty geometric proofs will be given. These "proofs" will then be analyzed in Chapters 2 and 4.

Among these propositions there are some whose falsity will be immediately apparent to the reader, for instance, "A right angle is equal to an obtuse angle." Here our task is to discover the mistake in the proof. Such proofs of deliberately incorrect assertions have been known from ancient times as "sophisms."

In other examples the reader will not know in advance whether the assertion proved is true or false if he has not come across it before. Here our task is more complicated; we must discover that the proof is unsound and whether or not the assertion is erroneous. To discover only the first is not sufficient; it is in fact possible to base a correct assertion on faulty arguments. For instance, from the incorrect equation $3 + 5 = 12$ it is possible to deduce correctly that $3 + 5$ is an even number.

Finally, examples will be given of proofs whose invalidity stems from the fact that the proposition asserted has nothing to do with the given data. How this may come about I shall attempt to explain by an example which is remote from geometry and science in general.

The following facetious problem is well-known: "A steamer is situated at latitude 42° 15′ N. and longitude 17° 32′ W. [The figures are taken at random; usually further data is added which complicates the conditions.] How old is the captain?" For our purpose let us alter the question of the problem somewhat. "Is the assertion correct that the captain is more than 45 years old?" It is clear to everyone that it is impossible to draw such a conclusion from the data given in the conditions of the problem, and that any attempt to prove the assertion concerning the age of the captain is destined to end in failure. Moreover, it is possible to *prove* that it is impossible to prove this assertion. Actually the steamship company, about which we learn nothing from the data of the problem, may chart a course which passes through the geographic point indicated and assign to the voyage a captain of this or that age, assuming that the company has captains of any age available for such trips.

In other words, it is possible to assume that the captain is younger than 45 years old without in any way contradicting the data concerning the latitude and longitude. It is another matter if the conditions of the problem contain other data as well, such as the name of the ship and the date on which it passes through the point indicated; one might then hope to establish the identity of the captain and then his age by using the ship's log. Thus, there exist assertions whose validity may or may not be proved, depending on the data which we have at our disposal for obtaining the proof.

Returning more nearly to our subject, let us ask, "Is it true that the sum of the angles of any triangle is equal to two right angles?" Every schoolboy who has studied the chapter on parallel straight lines in a geometry textbook is acquainted with the proof of this important theorem, but few know its history, which goes back 2,000 years. The proof is based on the properties of angles formed by a line intersecting parallel straight lines, and these properties are based in turn on the so-called "parallel postulate": *Only* one straight line can be drawn parallel to a given line through a given point not on this line.[1] Since the time of Euclid, efforts have been made to turn this axiom into a theorem, that is, to prove it only on the basis of the other axioms and on assertions which both in Euclid and in our textbooks do not depend on the parallel postulate. But we cannot "prove" this axiom like a theorem; all such attempts have been unsuccessful and it has been found only that the parallel postulate may be replaced by *equivalent* ones in many different ways. In particular, if we take as an axiom one of the properties of the angles formed by a pair of parallel straight lines and a third line intersecting them, or the theorem about the sum of the angles of a triangle, then the parallel postulate becomes a theorem.

It was not until the eighteen-twenties that the Russian mathematician Nikolai Lobachevskii (1792–1856) discovered the cause for the failure of all attempts to prove the parallel postulate. He constructed an extensive and profound theory of geometry, of which I shall not attempt to give here even the remotest idea. Among other things, this theory shows that it is impossible to prove the parallel postulate, which had been attempted by many scholars up to the time of Lobachevskii (and during his lifetime).

However complex the theory of Lobachevskii, and on the other hand, however naive the problem about the age of the captain, the "proof of the impossibility of the proof" is of the same nature in both problems. In these problems we are given certain data, and we desire to show that a certain conclusion cannot be logically deduced from the data. To do this we find concrete examples (called "models") in which all the given conditions are satisfied, but in some of which the conclusion in question is true and in others the conclusion is false. Applied to the parallel postulate this means that neither the truth nor the falsity of the parallel postulate follows from the other axioms of Euclidean geometry. We now know that any proof for the parallel postulate, or for any other postulate equivalent to it, must be invalid if it is based only on the propositions based on the other axioms. In Chapter 1 we shall give several simple examples of such faulty proofs.

[1] Note that the axiomatic nature of this proposition is based on the word "only." The fact that it is always possible to draw one parallel line can be proved earlier, if only on the basis of the theorem that two perpendiculars to the same straight line are parallel to each other.

1. Mistakes in Reasoning within the Grasp of the Beginner

We shall now proceed to give examples of faulty proofs, bearing in mind that their critical analysis will be postponed until Chapter 2. The reader has been forewarned that some of the drawings in this booklet contain distortions which are sometimes not at once apparent.

EXAMPLE 1. *The square whose side is* 21 *cm. has the same area as a rectangle whose sides are* 34 *cm. and* 13 *cm.*

The square Q is divided into two rectangles whose dimensions are 13×21 and 8×21 (Fig. 1; "cm." will subsequently be

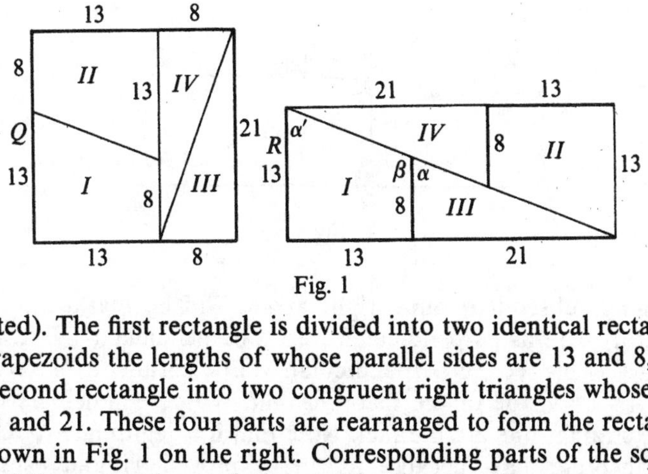

Fig. 1

omitted). The first rectangle is divided into two identical rectangular trapezoids the lengths of whose parallel sides are 13 and 8, and the second rectangle into two congruent right triangles whose legs are 8 and 21. These four parts are rearranged to form the rectangle R shown in Fig. 1 on the right. Corresponding parts of the square and of the rectangle are designated by the same Roman numerals.

Precisely speaking, we place the right triangle *III* next to the rectangular trapezoid *I* in such a way that the right angles at the common side of length 8 are adjacent; a right triangle is formed with legs 13 and $13 + 21 = 34$. An identical triangle is obtained from parts *II* and *IV*, and these two congruent right triangles are put together to form the rectangle R with sides 13 and 34.

The area of this rectangle is equal to

$$34 \times 13 = 442,$$

while the area of the square Q, which is made up of the same parts, is

$$21 \times 21 = 441.$$

Where does the extra square centimeter come from? We suggest that the reader carry out an experiment. Cut the square Q out of paper (preferably paper ruled in squares), taking the side of one square to represent 1 cm., dissect Q into four parts, carefully observing the dimensions indicated, and rearrange these parts to give the rectangle R.

EXAMPLE 2. *A proof of the parallel postulate.*

Given a straight line AB and a point C outside it, prove that through the point C only one straight line parallel to AB can be drawn. Using a familiar construction, drop a perpendicular from the point C to the straight line AB (Fig. 2; in this figure and

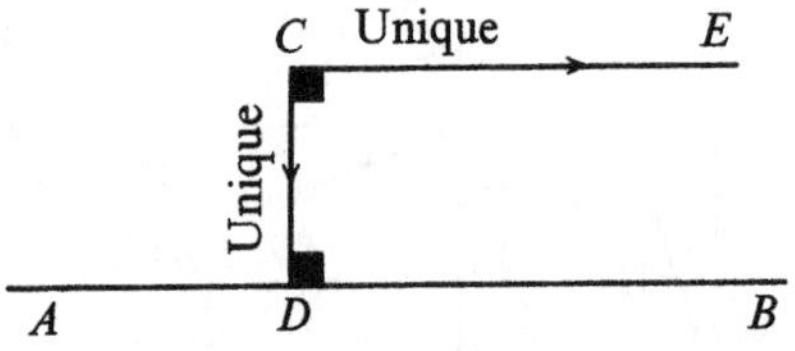

Fig. 2

in many subsequent ones, right angles will be marked by solid squares). To this perpendicular erect a perpendicular CE from the point C. This second perpendicular will be parallel to the straight line AB by virtue of the theorem that two perpendiculars to the same straight line are parallel. Note that it is legitimate to refer to this theorem here, because it can be proved without using the parallel postulate. But it is possible to drop only *one* perpendicular from a given point to a given straight line, and it is possible to erect only *one* perpendicular to a straight line from a point lying on it; both these facts can be proved without using the parallel postulate. Therefore, the parallel line CE obtained is *unique.*

EXAMPLE 3. *If two parallel straight lines are intersected by a third line, the sum of the interior angles lying on the same side of the third line is equal to* 180° (*proof not based on the parallel postulate*).

Say $AB \parallel CD$, and let the line EF intersect these two lines (Fig. 3). The interior angles are designated by numbers in the drawing.

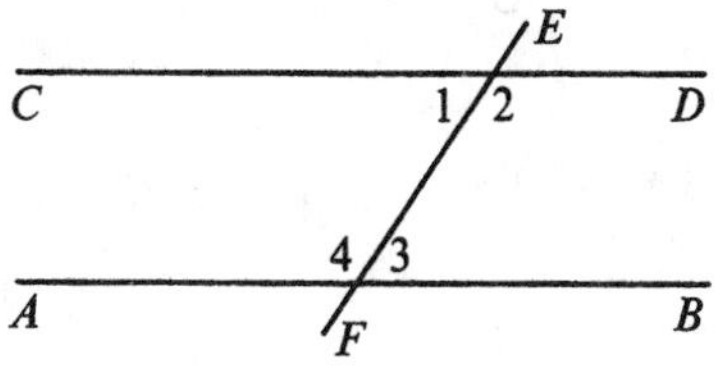

Fig. 3

Three cases are possible:[1]

1. The sum of the interior angles lying on the same side of the line EF is $> 180°$.
2. The sum of the interior angles on the same side of the line EF is $< 180°$.
3. The sum of the interior angles on the same side of the line EF is $= 180°$.

In the first case we have

$$\angle 1 + \angle 4 > 180°, \quad \angle 2 + \angle 3 > 180°;$$

therefore,

$$\angle 1 + \angle 2 + \angle 3 + \angle 4 > 360°.$$

But the sum of the four interior angles is equal to two straight angles, that is, 360°. This contradiction shows that the first assumption must be discarded. For the same reason we must also abandon the second assumption, as it would lead to the conclusion that the sum of the four interior angles is less than 360°. The third assumption is the only possible one (it does not lead to a contradiction); this proves the theorem.

[1] Here and subsequently, when talking about possible assumptions or possible cases, we do not by any means assert that they are all actually possible under the conditions of the given example. On the contrary, time and again it will happen that what is at first assumed to be a possible case later turns out to be spurious—contrary to the conditions or to what is taken as established; this often happens in indirect proofs. Thus, we are talking throughout about so-called "*a priori* possibilities," that is, about possibilities which present themselves beforehand, prior to taking into account the other conditions of the problem.

EXAMPLE 4. *The sum of the angles of a triangle is equal to* 180° (*proof not based on the parallel postulate*).

Divide the arbitrary triangle ABC into two triangles by means of a line segment drawn from the vertex, and denote the angles by numbers as in Fig. 4. Let x be the sum of the angles of a triangle, unknown as yet; then

$$\angle 1 + \angle 2 + \angle 6 = x,$$
$$\angle 3 + \angle 4 + \angle 5 = x.$$

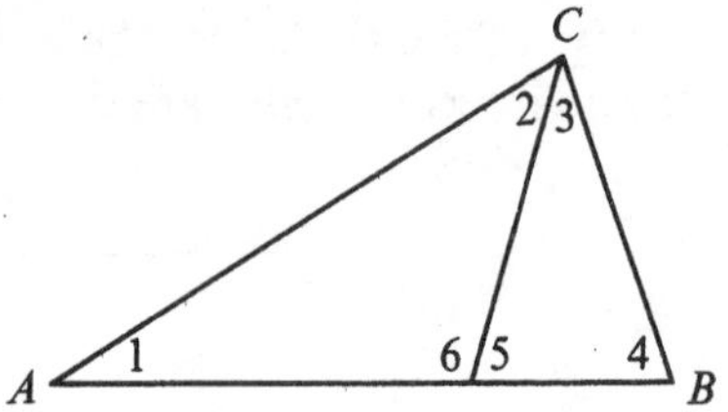

Fig. 4

Adding these two equalities, we obtain

$$\angle 1 + \angle 2 + \angle 3 + \angle 4 + \angle 5 + \angle 6 = 2x.$$

But the sum $\angle 1 + \angle 2 + \angle 3 + \angle 4$ is the sum of the angles of the triangle ABC, that is, it is also x; and the angles 5 and 6, being adjacent angles, have a sum equal to 180°. Thus, for finding x we have the equation $x + 180° = 2x$, from which it follows that $x = 180°$.

EXAMPLE 5. *There exists a triangle the sum of whose angles is equal to* 180° (*proof not based on the parallel postulate*).

We shall begin with a historical note. In the eighteenth and at the beginning of the nineteenth century some mathematicians attempted to show that it is possible to speak about the sum of the angles of a triangle without referring to the parallel postulate. It was established that the sum of the angles of a triangle cannot be greater than 180°. There remained three possibilities: (1) for all triangles this sum is equal to 180°, (2) for all triangles it is less than 180°, (3) it is sometimes equal to and sometimes less than 180°. It was subsequently found that the third of these possibilities can be excluded. Efforts were then concentrated on obtaining at least one example of a triangle the sum of whose angles is equal to 180°. We shall now describe one attempt to achieve this; if it should succeed, the parallel postulate would become superfluous.

As the sum of the angles of a triangle does not exceed 180°, let the triangle ABC (see Fig. 4) be a triangle the sum of whose angles is greatest; we shall designate this sum by α. If there are several such triangles, we shall take one of them at random. Thus, the sum of the angles of any other triangle will not exceed α; therefore, retaining the notation of Fig. 4, we have

$$\angle 1 + \angle 2 + \angle 6 \leq \alpha, \quad \angle 3 + \angle 4 + \angle 5 \leq \alpha.$$

From this we obtain $\angle 1 + \angle 2 + \angle 3 + \angle 4 + \angle 5 + \angle 6 \leq 2\alpha$; but by assumption $\angle 1 + \angle 2 + \angle 3 + \angle 4 = \alpha$, and moreover, $\angle 5 + \angle 6 = 180°$; consequently, $\alpha + 180° \leq 2\alpha$, $\alpha \geq 180°$. And inasmuch as α cannot be greater than 180°, we must have $\alpha = 180°$ this is, the sum of the angles of triangle ABC is 180°.

EXAMPLE 6. *All triangles are isosceles.*

Let ABC be an arbitrary triangle (Fig. 5, 6, or 7); construct the bisector of the angle C and the perpendicular bisector of the side AB. We shall consider the different relative positions of these lines.

Case 1. *The bisector of C and the perpendicular bisector of AB do not intersect*; they are either parallel or they coincide. The bisector of $\angle C$ will then be perpendicular to AB; that is, it will coincide with the altitude. Then triangle ABC is isosceles ($CA = CB$).

Case 2. *The bisector of* $\angle C$ *and the perpendicular bisector of AB intersect inside the triangle ABC*, say at the point N (Fig. 5). Since

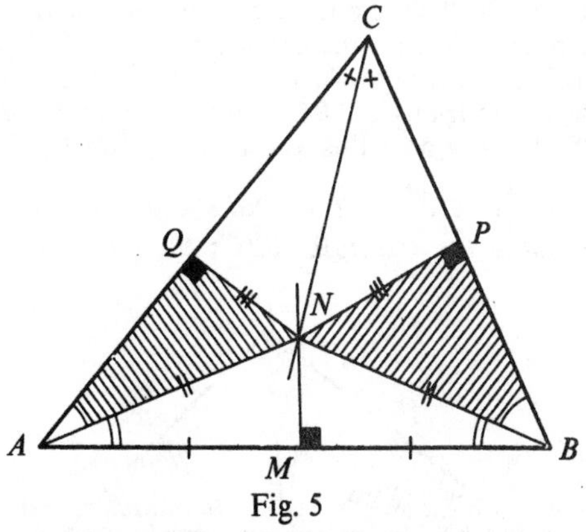

Fig. 5

this point is equidistant from the sides of the angle ACB, the perpendiculars NP and NQ to CB and CA, respectively, will be equal. But the point N is also equidistant from the end points of the line segment AB; that is, $NB = NA$. Then the right triangles NPB and NQA are congruent (leg-hypotenuse); hence, $\angle NAQ = \angle NBP$.

Adding to these equal angles the angles NAB and NBA, which are equal to each other since they are base angles of the isosceles triangle ANB, we obtain $\angle CAB = \angle CBA$; therefore, the triangle ABC is isosceles.

Case 3. *The bisector of $\angle C$ and the perpendicular bisector of AB intersect on AB*, that is, at the mid-point M of AB. This means that the median and the angle bisector from vertex C coincide; it follows that the triangle is isosceles.

Note. The reader is warned against a possible mistake. It is well known that the median and the angle bisector from the vertex opposite the base of an isosceles triangle coincide. We are referring here not to this, but to the converse assertion: "If the median and the angle bisector from the same vertex of a triangle coincide, the triangle is isosceles." This converse theorem is also true, but the reader may find it difficult to prove; we shall, therefore, indicate one possible method. In the triangle ABC suppose CM is both the median and the angle bisector. Dropping the perpendiculars MP and MQ to the sides CB and CA from the point M, we obtain the congruent right triangles MPB and MQA. Fig. 5 may be used, taking the points M and N to coincide; the straight line MN then becomes superfluous. Then from the equality of the angles MBP and MAQ we conclude that the triangle ABC is isosceles. This reasoning will be incomplete if we do not show that the points P and Q fall on the sides CB and CA themselves, and not on their extensions. One of these points might fall on the corresponding extension if either the angle A or the angle B were obtuse. Say, for instance, the angle B is obtuse so that the point P lies on the extension of CB. As before, we obtain $\angle MAQ = \angle MBP$; but this now leads to a contradiction, as the first of these angles is interior to the triangle ABC, while the second is exterior and not adjacent to the first. (Why does this lead to a contradiction?)

Case 4a. *The bisector of $\angle C$ and the perpendicular bisector of AB intersect outside the triangle ABC; the perpendiculars dropped*

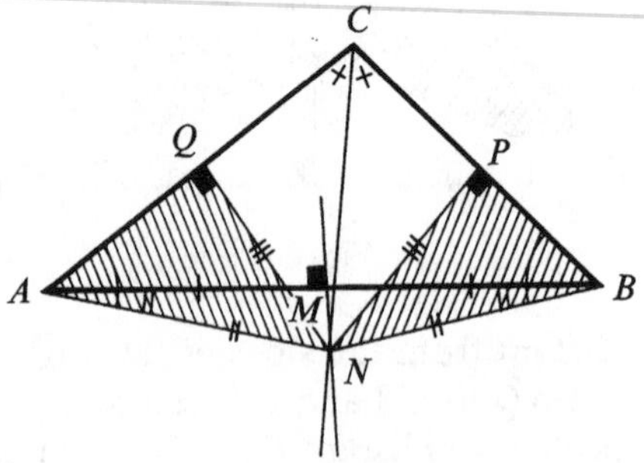

Fig. 6

from their point of intersection N to the sides CB and CA fall on these sides and not on their extensions (Fig. 6). As before, we obtain the congruent triangles NPB and NQA and the isosceles triangle ANB. The angles at the base AB of the triangle ABC are now equal, being the difference (not the sum, as in case (2)) of corresponding equal angles.

Case 4b. *The bisector of $\angle C$ and the perpendicular bisector of AB intersect outside the triangle ABC; the perpendiculars dropped from their point of intersection N to the sides CB and CA fall on their extensions* (Fig. 7). The same constructions and reasoning lead to the conclusion that the exterior angles at vertices A and B of the triangle ABC are equal. From this it follows immediately that the interior angles at A and B are equal; consequently, $CA = CB$.

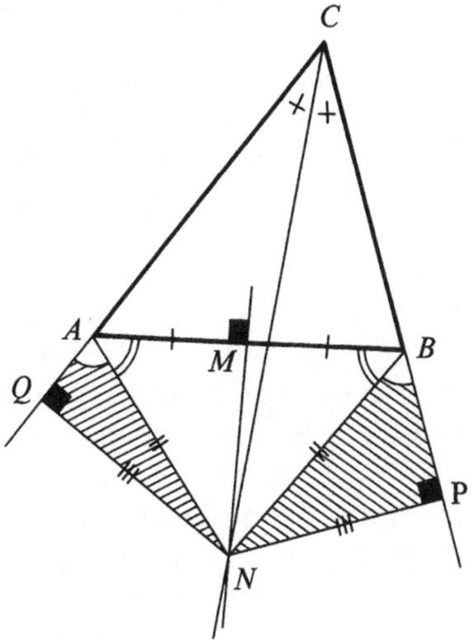

Fig. 7

EXAMPLE 7. *A right angle is equal to an obtuse angle.*

From the end points of the line segment AB (Fig. 8 or 9) draw two equal line segments AC and BD, lying on the same side of AB and forming with it the right angle DBA and the obtuse angle CAB; we shall now prove that these two angles are equal. By joining C and D we obtain the quadrilateral $ABDC$, whose sides AC and BD are clearly not parallel; the same applies to the sides AB and CD, for otherwise $ABDC$ would be an isosceles trapezoid with unequal angles at the base AB. Construct the perpendicular bisectors of each of the line segments AB and CD. Since the line segments AB and CD are not parallel, their perpendicular bisectors will not be parallel either and will not coincide, but will intersect at some point N.

Let us examine the possible cases.

Case 1. *The point N lies "above" the straight line AB*, strictly speaking, on the same side of AB as the quadrilateral $ABDC$ (see Fig. 8, where the point N is situated in the interior of the quadri-

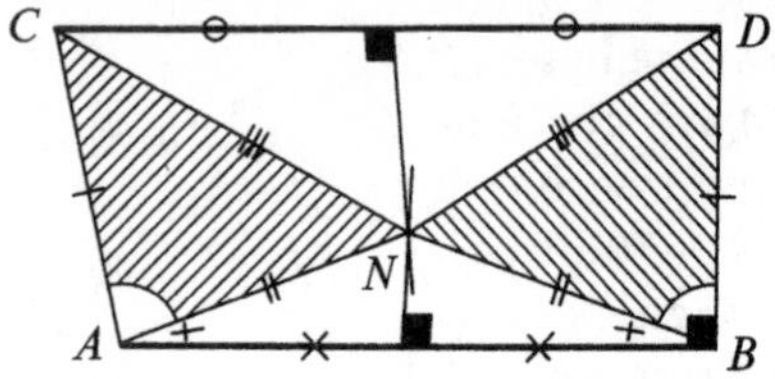

Fig. 8

lateral). Join this point with all the vertices of the quadrilateral. Since it is equidistant from the end points of the line segment AB and likewise from the end points of CD, the triangles NAC and NBD are congruent because the three pairs of corresponding sides are equal. From this it follows that $\angle NAC = \angle NBD$. Adding the angle NAB to the first angle and the angle NBA to the second and taking into account that $\angle NAB = \angle NBA$ by the properties of isosceles triangles, we arrive at the equality $\angle CAB = \angle DBA$.

Case 2. *The point N lies on AB*; that is, it is the mid-point of the line segment AB. The preceding proof can be simplified; the equality $\angle CAB = \angle DBA$ follows at once from the congruence of the triangles NAC and NBD.

Case 3. *The point N lies "below" AB*; that is, it does not lie on the same side of the line AB as the quadrilateral $ABDC$ (Fig. 9).

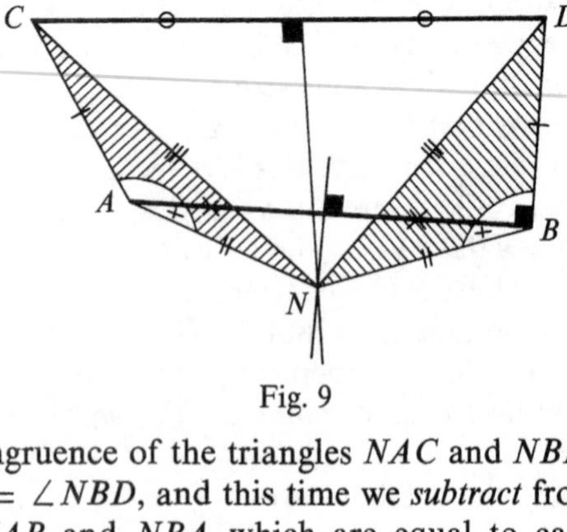

Fig. 9

From the congruence of the triangles NAC and NBD we again obtain $\angle NAC = \angle NBD$, and this time we *subtract* from these angles the angles NAB and NBA, which are equal to each other, again obtaining $\angle CAB = \angle DBA$.

EXAMPLE 8. *If two sides and the angle opposite one of them in one triangle are equal to the corresponding parts of another triangle, then these triangles are congruent.*

In the triangles ABC and $A_1B_1C_1$ (Fig. 10, 11, or 12) suppose we are given that

$$AB = A_1B_1,\ AC = A_1C_1, \text{ and } \angle C = \angle C_1;$$

we shall prove these triangles to be congruent. For this purpose we make use of a method widely used for the proof of the congruence of triangles with equal corresponding sides. Place triangle $A_1B_1C_1$ beside triangle ABC in such a way that the corresponding end points of the equal sides AB and A_1B_1 opposite the angles assumed equal coincide. The triangle $A_1B_1C_1$ will then occupy the position ABC_2. Joining the points C and C_2, let us examine three possible cases.

Case 1. *The straight line CC_2 intersects the side AB at a point lying between A and B* (Fig. 10). The triangle ACC_2 is isosceles;

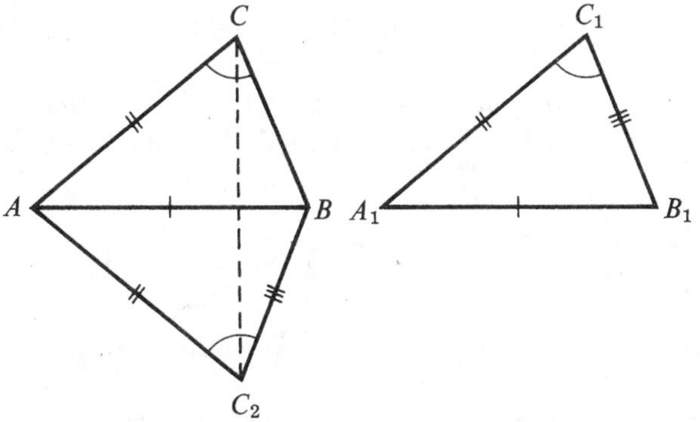

Fig. 10

consequently, $\angle ACC_2 = \angle AC_2C$. If we subtract these equal angles from the angles ACB and AC_2B, respectively, which are equal by hypothesis, we obtain

$$\angle BCC_2 = \angle BC_2C.$$

This last equality shows that triangle CBC_2 is also isosceles with $CB = C_2B$; therefore,

$$CB = C_2B = C_1B_1,$$

and the triangles ABC and $A_1B_1C_1$ are congruent (side-side-side).

Case 2. *The straight line* CC_2 *intersects the extension of* AB *beyond* B (Fig. 11). The reasoning is the same as before, except that

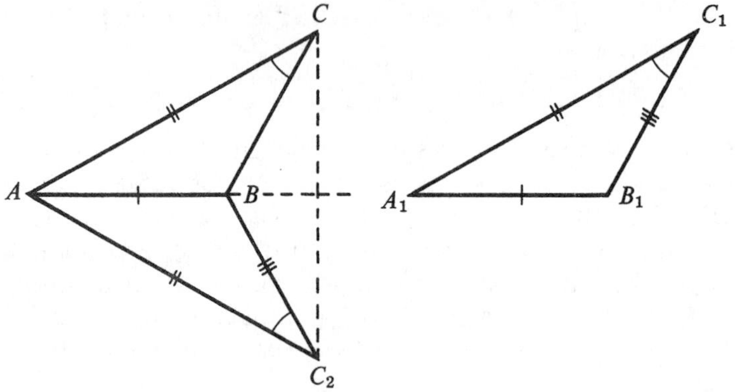

Fig. 11

this time the subtraction is performed in a different order; we subtract the equal angles ACB and AC_2B from the equal angles ACC_2 and AC_2C.

Case 3. *The straight line* CC_2 *intersects the extension of the side* BA *beyond* A (Fig. 12). The reasoning is the same as in case 1, but

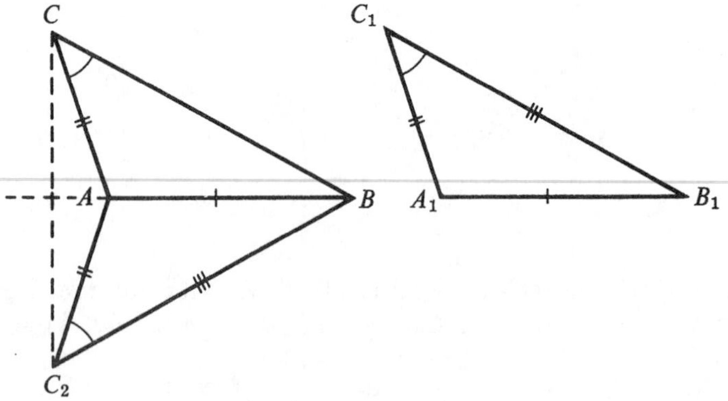

Fig. 12

addition takes the place of subtraction; to the equal angles ACC_2 and AC_2C we add the equal angles ACB and AC_2B.

EXAMPLE 9. *A rectangle which is inscribed in a square is also a square.*[1]

More precisely, if the rectangle $MNPQ$ (Fig. 13) is inscribed in the square $ABCD$ in such a way that one of the vertices of the rectangle lies on each of the sides of the square, the rectangle will also be a square. In the figure, M lies on AB, N on BC, P on CD, and Q on DA.

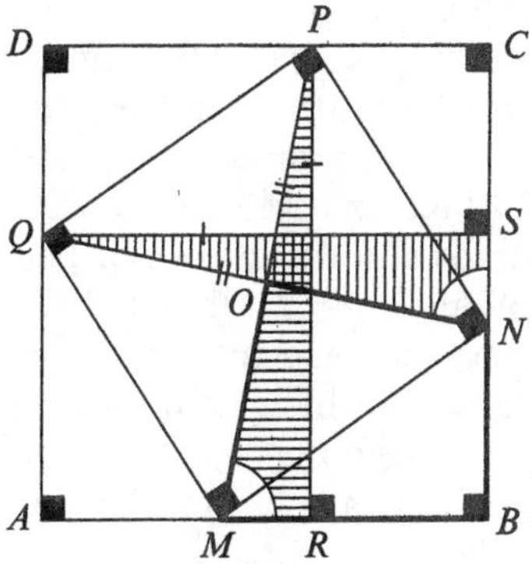

Fig. 13

To prove this, drop perpendiculars PR and QS from P and Q to AB and BC, respectively. These perpendiculars are equal to each other, for each is equal to the side of the square $ABCD$. They are the legs of the triangles PRM and QSN, whose hypotenuses, as the diagonals of the rectangle $MNPQ$, are also equal to each other. From this it follows that the triangles shaded in the figure are congruent, and hence that

$$\angle PMR = \angle QNS.$$

We now examine the quadrilateral $MBNO$ drawn in heavy lines in the figure, where O is the intersection point of the diagonals of the rectangle $MNPQ$. Its exterior angle at the vertex N is equal to the interior angle at the vertex M, so that the two interior angles at the vertices M and N are supplementary. The interior angles at the vertices B and O must also be supplementary, but one of them ($\angle B$) is a right angle, and, consequently, the other ($\angle O$) is also a right angle. Hence, the diagonals of the rectangle $MNPQ$ are perpendicular to each other. But this property of perpendicularity of the diagonals is a property which distinguishes squares from other rectangles.

Our proof is completed.

[1] We hope that no reader will feel that there is a contradiction in the combination of words "A rectangle . . . is . . . a square." Of course, not all rectangles are squares, but some of them are.

EXAMPLE 10. *Two lines, exactly one of which is perpendicular to a third line, do not intersect.*

This is a variation of an ancient sophism which has come down to us from the Greek mathematician Proclus (5th century A.D.).[1] Let us state the content of our assertion more precisely: At the points A and B of a straight line AB (Fig. 14) draw two half-lines lying on the same side of the line AB (in order to stress that these are in fact half-lines, they are marked with arrows in the drawing). Let AQ form an acute angle BAQ with AB, and BP be perpendicular to AB; we shall prove that these half-lines do not intersect.

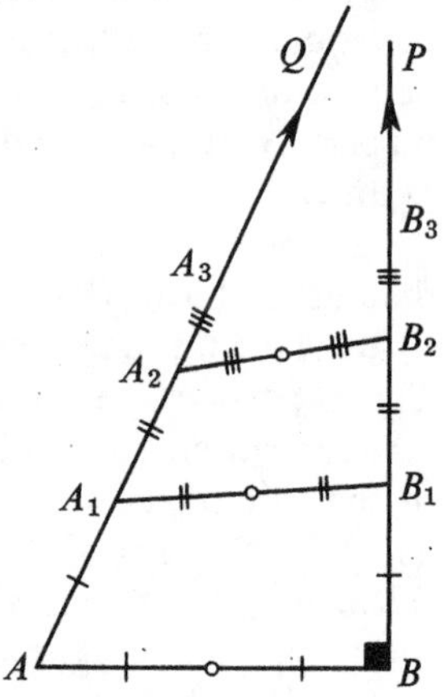

Fig. 14

Bisect the line segment AB and on each of the half-lines AQ and BP lay off $\frac{1}{2}AB$; thus, $AA_1 = BB_1 = \frac{1}{2}AB$. The perpendicular half-line and the oblique half-line cannot intersect anywhere along AA_1 and BB_1; that is, the segments AA_1 and BB_1 cannot have a common point. In fact, if a common point K did exist, we would obtain a triangle AKB in which the sum of two sides $AK + KB$ is less than or equal to the length of the third side AB, and that is not possible. Joining the points A_1 and B_1, we repeat the previous construction; on each of the half-lines AQ and BP and in the direction of the half-lines, lay off segments equal to $\frac{1}{2}A_1B_1$ from the points A_1 and B_1. We thereby obtain $A_1A_2 = B_1B_2 = \frac{1}{2}A_1B_1$. By the reasons advanced above, the line segments A_1A_2 and B_1B_2 cannot have a common point, and, in particular, A_2 cannot coincide with B_2. That being the case, we bisect the line A_2B_2 and lay off $A_2A_3 = B_2B_3 = \frac{1}{2}A_2B_2$, and so on. It must be stressed particularly that the equal distances $A_nA_{n+1} = \frac{1}{2}A_nB_n$ are laid off in the direction of the half-line in question each time. The process will continue indefinitely. It could be stopped if the segment A_nB_n were to disappear, that is, if for some number n the points A_n and B_n were to coincide; but, as we have seen, that is impossible. Besides, the impossibility of such a coincidence is clear directly from the fact that we would then obtain a right triangle for which the

[1] For an exposition of this sophism according to Proclus, see R. Bonola, *Non-Euclidean Geometry* (New York: Dover, 1955), pp. 5–6.

hypotenuse AA_n is equal to the leg BB_n. Thus, at no step of this infinite process does the intersection of the perpendicular half-line with the oblique half-line occur; therefore, it does not occur at all.

We have just seen a series of arguments which at times seem no less convincing than a proof from a geometry textbook. Sometimes the reasoning was directed at proving obvious absurdities; at other times the falsity of what was being proved was not immediately apparent, but the reader was told beforehand that each example did contain a mistake. The time has now come to uncover these mistakes.

Before proceeding to Chapter 2 for the analysis of all the examples given so far, we urge the reader to attempt to find the mistake in each example independently. It may be that not everyone will always successfully complete every example; nevertheless, one's own reflection about any of the examples prepares the ground for reading the analysis of the example in Chapter 2. If, on the other hand, the reader is successful, he will probably want to compare his interpretation with that given in Chapter 2. In suggesting this work, in which most readers will be inexperienced, I think it useful to give some preliminary hints and advice.

1. To refute an invalid geometric proof means to find a logical mistake in it. The difficulty lies in the fact that such a proof is correct almost everywhere, but always contains a gap at some point, and that gap has to be discovered.

2. In criticizing a proof it is often pointed out that it is carried out on the basis of "an incorrect drawing." This is not a very good criticism; in any case, it is not possible to limit oneself to this. When we say that drawing A is incorrect and should be replaced by drawing B, we often mask the following state of affairs: Not all possible cases are taken into consideration in the proof (and *that* is the logical mistake!). Conclusions valid for those cases which are considered and depicted in drawing A *alone* may turn out to be invalid for the other cases (drawing B). The source of the mistake is thus not in the drawing, but in the incomplete determination of the possible cases.

3. If the case depicted in drawing A leads to an absurd conclusion, it is sufficient to show that such a result is not obtained on the basis of drawing B, in order to prove indirectly the impossibility of the case A. It is also desirable, but not indispensable, to obtain

a direct proof of the fact that the hypotheses of the theorem lead necessarily to the case *B*. Examples of such proofs appear in Chapter 2.

4. Although a drawing cannot in itself reveal either the correctness or the falsity of an assertion, it is nevertheless recommended that all drawings be made as accurate as possible (by means of instruments). Where we are dealing with an obvious sophism, it is useful to make a drawing which will stress sharply the absurdity of the conclusion, for instance, in Example 7 by depicting an obtuse angle which is close to 180°, in Example 10 by drawing the two half-lines in such a way that they intersect within the limits of the drawing, and so on. Such a drawing can provide a clue for finding the mistake.

5. In some cases the mistake is in no way connected with the drawing, but consists, for instance, in that the proof is given (correctly) not for the assertion which one has set out to prove, but for an assertion related to it. Here the author of the proof has either not noticed the substitution himself, or else counts on others' not noticing it.

6. If it is not known whether or not the proposition being proved is true, it is best, though not obligatory, to begin by clearing up this question. It should be kept in mind that the assertion will have been refuted if even one example is constructed which contradicts it.

The reader will understand more clearly the significance of these hints after he has carried out independently the work suggested and read the succeeding chapter. I therefore recommend referring to these hints while reading Chapter 2 and considering them again after reading that chapter.

2. Analysis of the Examples Given in Chapter 1

EXAMPLE 1. In asserting that it is possible to rearrange the parts *I, II, III,* and *IV* of the square to obtain the rectangle we rely upon what seems to be obvious, or upon the evidence of a crude experiment, if we cut the pieces out of paper. On what basis can we conclude that a triangle is formed if the figures *I* and *III* (or, what amounts to the same thing, *II* and *IV*) are placed side by side, that is, that the oblique lateral side of trapezoid *I* and the hypotenuse of triangle *III* form a continuous straight line which is not bent at their common point? Of course the fact that we are not able to see such a bend in a drawing, or that we do not observe it when we perform the cutting experiment cannot be used as an argument. Even aside from the imperfection of our visual impressions, we note that they have to do not with geometric figures but with physical models of these figures, and therefore are of no use for rigorous geometric proofs.[1]

In order to recognize that the whole proof is unsound, it is sufficient to discover this gap. We may even refuse to discuss the matter further until this gap has been filled. However, we shall not follow this course but shall instead attempt to clear up the question of the bend completely.

If we could prove, for instance, that the angles α and β in Fig. 1 are supplementary, or else that the angles α and α' (in the same figure) are equal, the absence of a bend would be confirmed and the validity of the proof established. Is this possible? Reasoning indirectly, the answer is negative, for a positive answer to this question would lead to the equality $441 = 442$.

However, it is possible to ascertain directly that the angles α and α' are not equal, and at the same time to determine which of

[1] In the history of mankind this was by no means understood at once. In excavations of an ancient Indian temple dating back to 1,000 B.C., some mathematical records were found, among them a geometric figure depicted on the temple wall. The drawing apparently had to do with a rule for finding the area of a circle; instead of a proof, the word "Behold!" was written alongside the figure.

them is the greater. The next few lines will be within the grasp of the reader if he knows even a little trigonometry. (Instead of trigonometry the properties of similar triangles could be employed.) From triangle *III* in Fig. 1 we find the tangent of the angle α:

$$\tan \alpha = \tfrac{21}{8}.$$

If in trapezoid *I* we drop the perpendicular (not shown in Fig. 1) from the vertex of the angle β to the longer base, a right triangle is formed with legs of length 13 and $13 - 8 = 5$. From this we obtain

$$\tan \alpha' = \tfrac{13}{5}.$$

But $\frac{21}{8} - \frac{13}{5} = \frac{1}{40}$; therefore, $\frac{21}{8} > \frac{13}{5}$, and $\tan \alpha > \tan \alpha'$. From this it follows that

$$\alpha > \alpha', \quad \alpha + \beta > 180°.$$

Now the picture is clear; the parts *I, II, III, IV* of the square may actually be placed into the rectangle, but they do not cover it completely; a gap remains in the form of a very thin parallelogram —a "crack" which lies along the diagonal of the rectangle. It is not surprising that we do not notice this crack; for over the total length of 36.4 . . . (cm.) it has an area of only 1 (cm.²), which is exactly the excess area of the rectangle *R* over the square *Q*. A reader who wishes to make the picture still more obvious might change the numerical data in Fig. 1, for instance, in the way shown in Fig. 15,

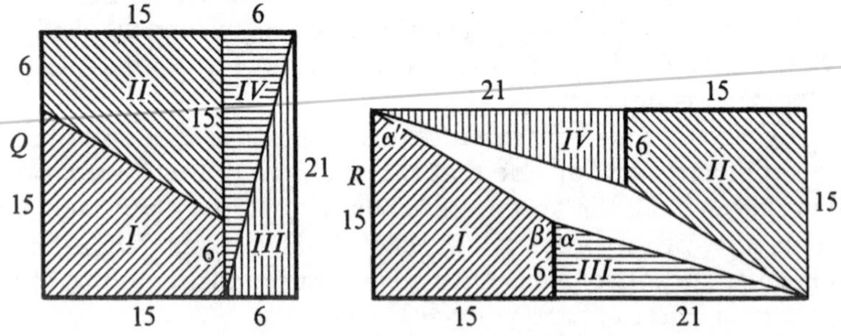

Fig. 15

where the "crack" has an area of 99 (cm.²), and the entire rectangle a total area of 540 (cm.²).

EXAMPLE 2. The mistake made here is of a type which was well known in classical logic, bearing the complicated Latin name *ignoratio elenchi*, which freely translated means "failure to understand what has been proved," or proving something other than what was required.

What, indeed, does the argument using Fig. 2 actually establish? It is shown only that a unique straight line is obtained if the parallel line is constructed by the method described there (by means of two perpendiculars). But are there not other ways to construct parallel lines? Yes, it is well known that other constructions exist which lead to the same end.

For instance, instead of base D of the perpendicular CD from C (see Fig. 2), we might take any other point D' on the straight line AB (Fig. 16), join it to C by the straight line $D'F$, and on the

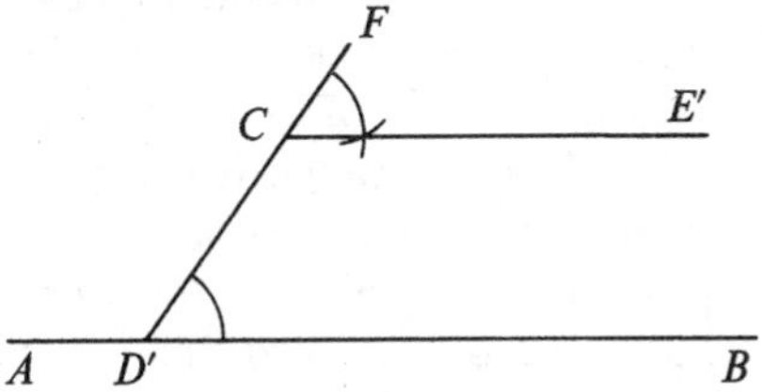

Fig. 16

half-line CF at the point C construct the angle FCE' equal to the angle $CD'B$ in such a way that the half-lines CE' and $D'B$ lie on the same side of FD'. On the basis of the theorem that straight lines making equal corresponding angles with a transversal are parallel, which can be proved without introducing the parallel postulate, it is possible to assert that the straight line CE' is parallel to AB. But where is the guarantee that the straight line CE in Fig. 2 coincides with CE' in Fig. 16? To assert that different constructions lead to the same straight line is to accept without proof that which we set out to prove.[1]

[1] In the geometry of Lobachevskii the straight lines CE and CE' do not coincide. Interested readers may wish to refer to N. Lobachevskii's *Geometrical Researches on the Theory of Parallels,* trans. by George B. Halsted, Open Court Publishing Co.

EXAMPLE 3. In considering just the three cases, we have made the following assumption: For any pair of parallel lines intersected by any third line, the sum of any pair of interior angles lying on one side of the third line will be always greater than, always equal to, or always less than 180°. Only at first glance will it appear that all possible cases have been exhausted; the possibility that the sum of the interior angles on one side of the third line is sometimes greater, sometimes less, and sometimes equal to 180° has been omitted. This assumption does not lead to any contradiction. For instance, the supposition that

$$\angle 1 + \angle 4 > 180^\circ \text{ and } \angle 2 + \angle 3 < 180^\circ$$

(see Fig. 3) may not in any way contradict the relationship

$$\angle 1 + \angle 2 + \angle 3 + \angle 4 = 360^\circ.$$

Note that without going into a detailed analysis of the proof it is possible to discover its lack of substance at the very outset by one simple observation—the proof does not use the fact that the straight lines AB and CD are parallel. If the proof were correct as it stands, the following theorem would have been proved: "If any pair of straight lines is intersected by a third line, the sum of the interior angles on one side of the third line is equal to 180°." But this is incorrect. It is precisely when the parallelism of the straight lines AB and CD is discarded that, as a rule, the fourth possibility, which was omitted in the fallacious proof, will materialize—the sum of the interior angles on one side of the intersecting line will be greater than 180°, and on the other side less than 180°.

EXAMPLE 4. We have become accustomed to accept the statement that the sum of the angles of a triangle is constant (180°) for all triangles, irrespective of their shape and dimensions. For this reason the majority of us do not protest against the statement, "We shall denote the sum of the angles of any triangle by 'x'." But actually when we set out to prove the theorem in question, nothing is known concerning the sum of the angles of the triangle, and there is no basis whatsoever for assuming that it is the same for all triangles. Of course, we could accept without proof the fact that the sum is the same and in that case the arguments advanced would indeed prove that this sum is equal to 180°. But this would merely mean that we had introduced another postulate in place of the parallel postulate.

EXAMPLE 5. In the history of mathematics several cases are known in which the very same mistake has been made; that is, with no basis it has been accepted that the terms of a given infinite set must necessarily include a greatest term (or a least term).

Now, it would not occur to anyone to seek the greatest term of the numbers

$$1, 2, 3, \ldots$$

of the sequence of natural numbers, as the absence of such a number is explained by the fact that in this case the numbers are increasing all the time and this sequence has no end. Moreover, the sequence of fractions whose numerator is one less than the denominator,

$$\tfrac{1}{2}, \tfrac{2}{3}, \tfrac{3}{4}, \tfrac{4}{5}, \ldots,$$

may also be continued indefinitely by repeatedly adding 1 to both the numerator and the denominator. As in the case of the first sequence, the numbers will increase, and there is no greatest term among them.

Closer to our problem is the following example taken from geometry: The interior angle of a regular polygon, equal to

$$\left[\frac{180(n-2)}{n}\right]^\circ$$

where n is the number of sides, is always less than 180°, but there is no regular polygon with a greatest interior angle.

The weak point in the proof under consideration is precisely the assumption that among all triangles whose angles we know do not add up to more than 180°, there exists a triangle for which this sum has a greatest value. This is an unproven assertion which we could accept as a new postulate in place of the parallel postulate.

By combining the results of the analysis of Examples 4 and 5 we conclude that it is possible to prove that the sum of the angles of a triangle is equal to 180° and at the same time render the parallel postulate superfluous if we accept without proof one of two assertions:

(1) The sum of the angles of all triangles is the same.

(2) There exists at least one triangle for which the sum of the angles is greatest.

EXAMPLE 6. Not all possible cases have been examined (in this connection it is useful to recall the footnote on page 7); in fact, no account has been taken of the possibility that one of the two perpendiculars NP and NQ falls on a side of the triangle ABC while the other falls on the extension of a side. (See Fig. 17, ignoring the

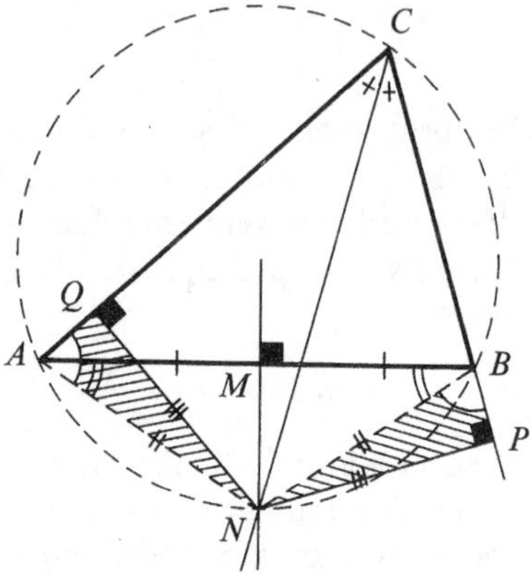

Fig. 17

circle for the present.) If this occurs, one of the angles at the base AB of the triangle ABC will be found to be the difference of two angles, while the other will be adjacent to the sum of the same two angles; from this, of course, no conclusion can be drawn regarding the equality of the angles at the base and of the lateral sides AC and BC. Establishing this gap in the proof is sufficient to discredit it. Moreover, if the given triangle is *not* isosceles, it may be asserted, arguing indirectly, that none of the cases considered (Fig. 5, 6, 7) will occur and that the only possible case (Fig. 17) has been omitted.[1]

We shall now give a direct proof that the parts of a nonisosceles tri-

[1] At first it may seem that the case for which the point N lies inside or on the base of the triangle and the points P and Q lie on different sides of AB has also been omitted. It is quite possible that a perpendicular dropped from a point inside the triangle to one of its sides should fall on the extension of the side; it is sufficient to consider an obtuse triangle. In the next paragraph we shall, however, establish that for a nonisosceles triangle the point of intersection of the perpendicular bisector of the base and the bisector of the opposite angle *must* lie outside the triangle. If the reader is acquainted with the theorem that the bisector of an angle of a triangle divides the opposite side into parts proportional to the two remaining sides, he might attempt to establish this property of the point of intersection by a different method.

angle are arranged precisely as shown in Fig. 17. For convenience, we suppose that $CA > CB$. We circumscribe a circle about the triangle ABC. From the property of inscribed angles, the bisector of angle C must pass through the mid-point N of the arc AB which angle C intercepts. But the perpendicular bisector of the chord AB must pass through the same mid-point N. Thus, the point of intersection of the bisector of $\angle C$ and the perpendicular bisector of AB falls on the circumscribed circle; that is, it lies *outside* the triangle ABC.

The perpendiculars dropped from N to sides CB and CA will fall on these sides or their extensions, depending on whether the angles NAC and NBC are acute or obtuse. Instead of these inscribed angles we shall examine the arcs which they intercept. Since we assumed $CA > CB$, we have $\overparen{CA} > \overparen{CB}$, and from $\overparen{AN} = \overparen{BN}$ it follows that $\overparen{CAN} > \overparen{CBN}$. This means that $\overparen{CAN}$ is greater than a semicircle and $\overparen{CBN}$ is less than a semicircle. Consequently, the angle CBN is obtuse and the angle CAN acute. The perpendicular NP, therefore, falls on the extension of CB, while the perpendicular NQ falls on the side AC itself. (As an exercise we suggest that the reader prove that the points P, M, and Q lie on a straight line.)

EXAMPLE 7. The proof appears at first to be convincing, as it creates the illusion that all essentially different cases have been examined[1]—the point N lies above, below, or on the straight line AB. However, the course of the proof does not depend *only* on the position of the point N. Notice that in case (3) the right angle ABD together with the acute angle ABN will always give an obtuse angle DBN; however, the obtuse angle CAB, when added to the acute angle NAB, may either again give an obtuse angle (Fig. 9) or else a reflex angle (greater than 180°, see Fig. 18), which changes the matter fundamentally.

Thus, case (3) must be divided into two subcases: the obtuse angle CAB and the triangle CAN lie (1) on the same side of the straight line AC (see Fig. 9), or (2) on opposite sides of it (see Fig. 18, which should first be examined ignoring the dotted lines). The first subcase, in which the angle CAB is a part of the angle CAN, has been examined and leads to the equality of the angles DBA and CAB. The second subcase, however, does not lead to this

[1] Two cases must be regarded as essentially different if a proof which is valid for the one case cannot be applied word for word to the other case.

result; as before, the right angle DBA is the *difference* of two angles ($\angle DBN$ and $\angle ABN$), but the obtuse angle CAB together with the

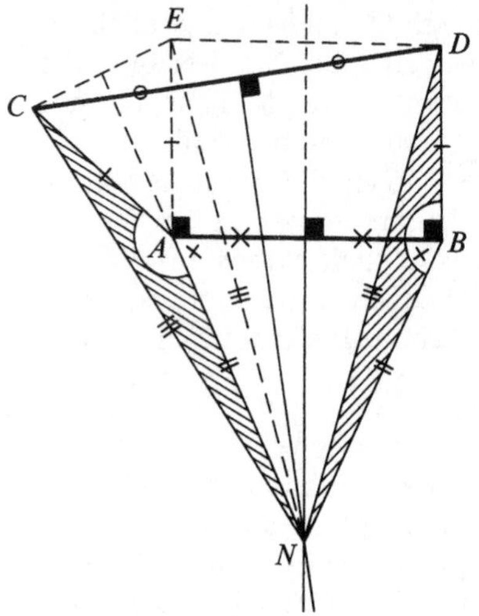

Fig. 18.

sum of the corresponding angles ($\angle CAN$ and $\angle BAN$) totals 360°. Arguing indirectly we can conclude that the second subcase is the only one possible.

We shall carry out an additional construction which will afford a clearer view of the arrangement of the parts of the figure. At the point A erect a perpendicular to AB (now the dotted lines in Fig. 18 come into the picture), and on it lay off the line segment AE equal and parallel to BD; evidently, we have $AE = AC$. Join E to the points D, N, and C. Since $ABDE$ is a rectangle, the perpendicular bisector of AB will also be the perpendicular bisector of ED; consequently, $NE = ND$, and hence $NE = NC$. Thus, each of the points A and N is equidistant from the end points of the line segment CE; consequently, the straight line AN is the perpendicular bisector of this line segment. The triangle DBN is transformed into the triangle EAN (oriented in the opposite direction) by a reflection in the perpendicular bisector of AB, and the triangle EAN is transformed in turn into the triangle CAN, with the same orientation as the triangle DBN, after a reflection in the straight line AN. Thus, CAN can be obtained from DBN simply by a rotation about the vertex N through $\angle BNA$, which is equal to $\angle EAC$, that is, the difference between the original obtuse and right angles.

EXAMPLE 8. Let us make certain first of all that the theorem is not true. For this purpose it is sufficient to furnish a counterexample, that is, a case in which the conditions of the theorem are fulfilled, but not the conclusion. We obtain such an example if we divide any isosceles triangle LMN ($LN = MN$, Fig. 19) into two parts by means

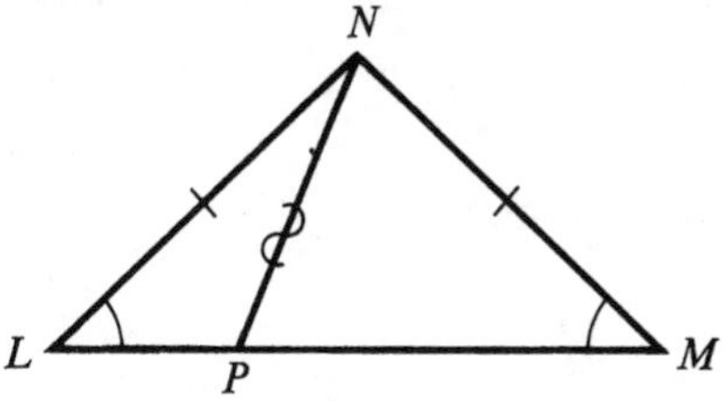

Fig. 19

of an oblique line segment NP from the vertex. In the triangles LNP and MNP so obtained, NP is a common side, and, moreover, $LN = MN$ and $\angle L = \angle M$. Thus, the conditions of the theorem are fulfilled, although the two triangles are, of course, not congruent as $LP \neq MP$.

But even if it is not known whether or not the theorem is true, it is possible to discover a gap in the proof. This gap lies in our omission of the cases in which the straight line CC_2 passes through one of the end points of the line segment AB, that is, in which the sides CA and C_2A or CB and C_2B are extensions of each other (using the notation of Figs. 10–12).

If the two *equal* sides AC and AC_2 lie on the same straight line (Fig. 20), the conclusion of the theorem is still correct; by joining

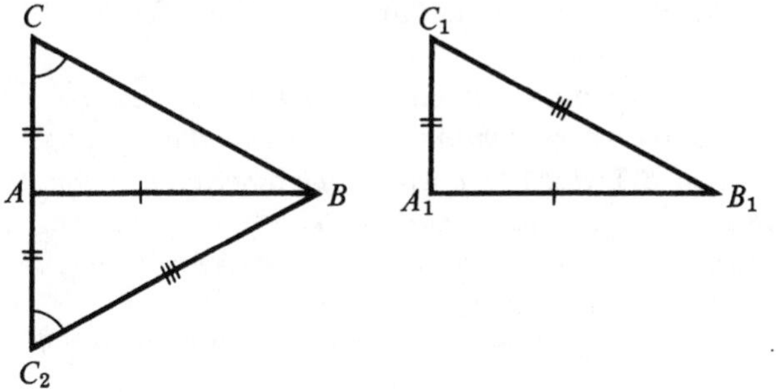

Fig. 20

the triangle $A_1B_1C_1$ to the triangle ABC, we obtain the triangle BCC_2, which is isosceles because angles C and C_2 are equal; consequently, $BC = BC_2 = B_1C_1$, and the triangles are congruent.

Let us add that this occurs only with right triangles; in the left of Fig. 20 the angles at the point A are equal and supplementary and, hence, right angles.[1]

A different picture is obtained if those sides, BC and BC_2, about which nothing is known from the hypotheses lie on a straight line (Fig. 21). Of course, we obtain the isosceles triangle ACC_2, but it is

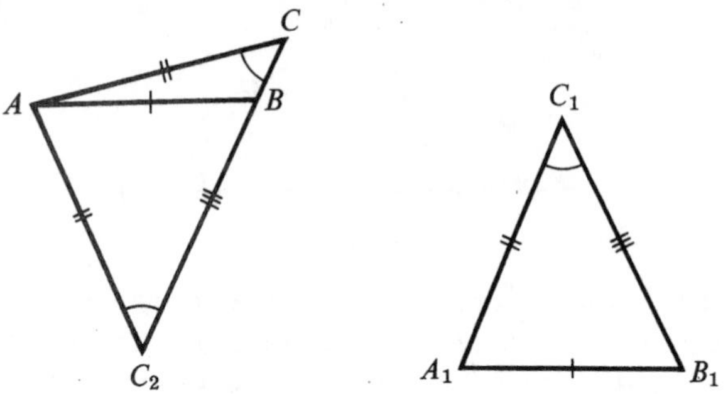

Fig. 21

impossible to draw from this any conclusions relating to the sides CB and C_2B. Furthermore, the reader will immediately recall the figure of an isosceles triangle (Fig. 19) divided into two unequal parts. Hence, we can conclude that the triangles ABC and $A_1B_1C_1$ are not congruent in this case unless the angles at the vertices B and B_1 are right angles.

Note 1. The foregoing arguments suggest ways in which our theorem might be "amended," that is, how it might be replaced by some related theorems which are valid. We offer two "amended" versions.

1. *If two sides and the angle opposite one of them in one triangle are equal to the corresponding parts of another triangle, then the angles (B and B_1) which lie opposite the other pair of equal sides will either be equal to each other and the triangles will be congruent* (*Fig.* 10–12, 20), *or else these angles will be supplementary* (*Fig.* 21), *and the triangles will not be congruent.*

2. *If two sides and the angle lying opposite the larger of them in*

[1] Note that this possibility of C, A, and C_2 lying on a straight line must also be anticipated in the usual proof for the congruence of triangles by three equal corresponding pairs of sides. There the corresponding sections of the proof can be carried out successfully, and it is thereby also found that the congruent triangles are right triangles.

one triangle are equal to the corresponding parts of another triangle, then these triangles are congruent.

The second statement excludes the case depicted in Fig. 21. Neither of the angles B and B_1 can be obtuse or even a right angle, as C or C_1 would not then lie opposite the larger side.

Note 2. The reader may at first be puzzled by the following assertion: The case of congruence of the two triangles can, in a certain sense, be made "as close as we like" to the case where they are not congruent. We may regard Fig. 21 as a "degenerate" case of Fig. 10, in which the point B has slid onto the line CC_2. In Fig. 10, B could lie as close as we like to CC_2, and $\triangle CBC_2$ could be as flattened as we like, but it would necessarily be isosceles, and the proof of congruence would be valid. But as soon as B gets *onto* CC_2, the equality of CB and C_2B can no longer be demonstrated, and, indeed, may not hold. We shall discuss briefly a point of view which will throw some light on this.

It is sometimes convenient to regard three points which lie on a straight line as the vertices of a "degenerate" triangle; if Q lies between P and R, the angles of the "triangle" will be $\angle P = 0°$, $\angle R = 0°$, $\angle Q = 180°$. The meaning of this terminology is clear; as long as the points lie "almost" on a straight line, they still determine a triangle with two extremely small angles, and the third angle is nearly an angle of 180°. The figure can be continuously deformed so that these three points come to be situated exactly on a straight line, and it is desirable to retain the previous terminology. Some theorems are valid for both "real" and degenerate triangles; for instance, the theorem that the sum of the angles of a triangle is equal to two right angles. There are, however, other theorems which are not valid for degenerate triangles. Among these, in particular, is the theorem stating that a triangle with two equal angles is isosceles. This theorem is valid if the equal angles differ from zero; but in the degenerate triangle PQR described above, it does not follow at all from the equality $\angle P = \angle R = 0°$ that $PQ = QR$, that is, that Q is the mid-point of PR; Q might lie anywhere on the line segment. This example is directly related to Example 8. As long as the point B in Fig. 10 or 11 does not lie on the straight line CC_2, then no matter how small the angles are at the side CC_2 in the triangle BCC_2, it follows from the equality of these angles that $CB = C_2B$. But if, as in Fig. 21, the triangle BCC_2 is degenerate, then both of the angles become zero, and the conclusion that the sides are equal and with it the conclusion of the theorem, lose their validity.

EXAMPLE 9. The assertion of the theorem is erroneous, as we may show by constructing a counterexample. It is sufficient to take the sides of the rectangle parallel to the diagonals of the square, and its vertices not bisecting the sides of the square. More precisely, from two opposite vertices, A and C, of the square lay off along its sides four equal segments $AM = AQ = CN = CP$ of any length not equal to $\frac{a}{2}$, where a is the length of a side of the square. Then $MNPQ$ is a rectangle, for its sides make angles of 45° with the sides of the square, and so are parallel to its diagonals. But these are perpendicular, and so the sides of $MNPQ$ are also perpendicular (Fig. 22).

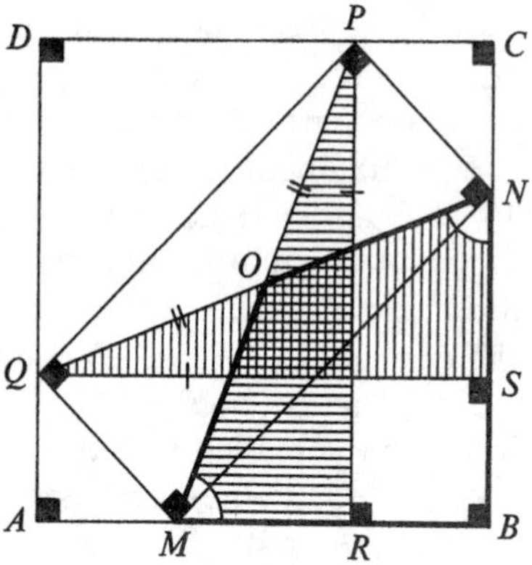

Fig. 22

Quite apart from this construction, we may see directly that the proof we gave in Example 9 was fallacious. Relying only on the appearance of Fig. 13, we assumed that of the two projections, P onto AB and Q onto BC, one lies inside and the other outside the quadrilaterial $MBNO$. We see from Fig. 22 (where we do not now assume any more than is marked on the figure) that this may not be so. Our original proof, with S and R both lying *inside* $MBNO$, would show merely that $\angle OMB = \angle ONB$. Let us carry the argument further. $\triangle MON$ is isosceles, being formed by the diagonals of a rectangle. So $\angle OMN = \angle ONM$, and, by subtraction, $\angle BMN = \angle BNM$. By considering $\triangle BMN$, we see that these angles are 45°, and so our rectangle is exactly of the type we constructed above. To complete the analysis, we must examine the case where both R and S fall outside $OMBN$, which turns out to lead to the same result as the case we have just considered, and the case where one of R and S coincide with a vertex of $OMBN$, which turns out to give a square. We leave the proofs to the reader.

To summarize, we may replace the false theorem by a more complicated theorem such as:

1. *If a rectangle is inscribed in a square in such a way that one of the sides of the rectangle is not parallel to either of the diagonals of the square, then the rectangle is a square;* or

2. *If a rectangle with unequal sides is inscribed in a square, the sides of the rectangle must be parallel to the diagonals of the square.*

EXAMPLE 10. The logical mistake is the same as that in Example 2: "failure to understand what has been proved." In other words, for the proposition to be proved we have substituted another proposition which actually holds, but from which the statement to be proved does not in any way follow. Let us consider once more the line of reasoning and simplify the task of discovering the mistake by replacing Fig. 14 by Fig. 23, in which the half-lines AQ

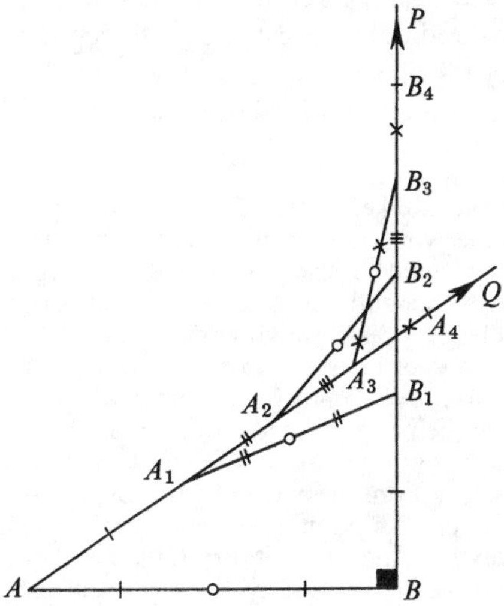

Fig. 23

and BP do intersect. Both figures have been made without any deliberate distortion. Let us indicate by AA_1, A_1A_2, A_2A_3, ... the first, second, third, ... segments on the oblique line AQ, and let us indicate by BB_1, B_1B_2, B_2B_3, ... the first, second, third, ... segments on the perpendicular BP. It must be taken as proved that (1) the process of laying off these segments may be continued indefinitely so that it is possible to obtain segments with arbitrarily high index, and that (2) segments with the *same index* do not intersect, that is, the first segment of the perpendicular has no common point with the first segment of the oblique line, the second segment of the perpendicular has no common point with the second segment of the oblique line, nor the hundredth with the hundredth, etc. But why should not segments with *different indices* intersect, say, the twentieth segment of the perpendicular with the twenty-fifth segment of the oblique line? For when we assert that the perpendicular and the oblique line do not intersect anywhere, we must prove that none of the segments of the perpendicular has a common point with *any* of the segments of the oblique line. And we must not be satisfied with proving instead that none of the segments of the perpendicular intersects the *corresponding* segment of the oblique line. If we turn to Fig. 23, in which the notation of Fig. 14 is preserved, we see by inspection that the second segment of the perpendicular and the fourth segment of the oblique line do intersect.[1] This sophism is noteworthy for the contrast between the elementary nature of the mistake and the difficulty of discovering it.

Note. As we have said before (see the footnote on page 16), we have used only the idea underlying the sophism given by Proclus. He examines two straight lines taken arbitrarily—in actual fact, two half-lines which do not lie on a straight line and which have different origins—and proves by means of the endless process of laying off segments described above, that these straight lines do not intersect. Proclus correctly characterizes the logical error contained in this sophistic argument when he says that the only thing proved is that the point of intersection cannot be found by the method of construction used, but this does not in any way signify that such a point does not exist. Judging from the account given by Bonola, however, it is not certain that Proclus penetrated more deeply into the geometric substance of the mistake; in any case the 19th-century Italian author is clearly mistaken when he states that the inaccessability of the point of intersection is brought about by the same reasons that give rise to the famous sophism about "Achilles and the tortoise." By this comparison Bonola means, of course, that the point of intersection, say K, of the half-lines AQ and BP is unattainable by the given construction simply because as n increases without bound, the points A_n and B_n approach K as their limit, but never reach it. In our version such an assumption is not possible, for from the equality $AA_n = BB_n$, which holds for arbitrary n, it would follow that $AK = BK$, that

[1] Knowing the angle A, it would be possible to calculate the indices of the intersecting segments by means of trigonometry.

is, that the hypotenuse is equal to one of the legs in triangle *ABK*. But this is impossible here and also in the construction of Proclus-Bonola except when the triangle *AKB* is isosceles. Thus, what happens generally is that segments with different indices intersect, and segments with the same index do not tend to a common limit.

CONCLUSIONS. The reader may ask the following: If mistakes in mathematical arguments are sometimes masked to such an extent that they can be discovered only after careful analysis, does mathematics really provide such a reliable foundation for the exact sciences (physics, engineering, and others) as we customarily believe?

Of course, no scientific method is a guarantee against faulty conclusions; the method in question must also be applied correctly. This only shows that one should study the sources of possible errors and be more exacting in substantiating one's assertions. In order to perceive the danger of committing an error which may pass undetected, we must turn to the history of our science.

This history has witnessed various mistakes in the works of mathematicians, but these mistakes have never stopped the progress of the science and have been exposed when a higher stage of development was reached. As an imposing example we may cite the previously mentioned history of the attempts made through many centuries to prove the parallel postulate. About this postulate Lobachevskii wrote in 1823, "It has not been possible to find a rigorous proof for this truth thus far. Such proofs as have been given can be called only explanations, and do not deserve to be dignified as mathematical proofs in the full sense of the term." Lobachevskii arrived at this conviction a few years before his outstanding discovery. From the new level of this discovery in the history of geometry it became obvious, first to Lobachevskii and subsequently to the whole mathematical world, that the most ingenious plans for proving the parallel postulate could never succeed.

3. Mistakes in Reasoning Connected with the Concept of Limit

The examples in this chapter are within the grasp of readers with some acquaintance with the simplest properties of circles, the concept of a limit, trigonometry, and some solid geometry.

EXAMPLE 11. *The circumferences of all circles are of equal length.* This ancient sophism is ascribed to the Greek philosopher Aristotle (fourth century B.C.), and for a reason which will soon become clear it is called "Aristotle's wheel."

Let us recall arithmetic problems in which we are given the length of the circumference of a wheel of a cart or car moving along a road, and we are to find the distance traveled, or vice versa. As the basis for the solution we take the seemingly obvious fact that for each complete turn the rolling wheel travels a distance equal to (the length of) its circumference. If, for instance, the circumference of the wheel is 2 meters and in rolling it has made 30 complete revolutions, the distance through which it has traveled will be 60 meters. Where the motion takes place in a straight line, and where no special accuracy is required, these calculations can be confirmed by experiment. The circumference of the wheel may be measured with a tape; the completion of a revolution of the wheel may be judged by marking one of the spokes of the wheel, or instead of this by attaching to the rim a band which leaves an imprint on the ground. (Many metering devices installed on different transport vehicles register the number of revolutions but indicate the distance, or, in combination with a clockwork mechanism, the velocity.) Of course, all these calculations are correct in practice only if the wheel turns "normally," that is, if it does not jump or slip; in the language of mechanics this is expressed by the phrase "the wheel rolls without slipping."

We shall now return to our sophism. Let us examine two concentric circles C and C_1 of different radii, which are rigidly attached to each other (Fig. 24). At the same time think of a physical model, two cylindrical rollers mounted on a common axis, which we shall take as horizontal, and rigidly attached to each other. (We could do with a cylindrical roller, part of which has the form of another cylindrical roller with the same axis but with a smaller radius; see the picture in Fig. 24). Draw the tangents MN and M_1N_1 to the

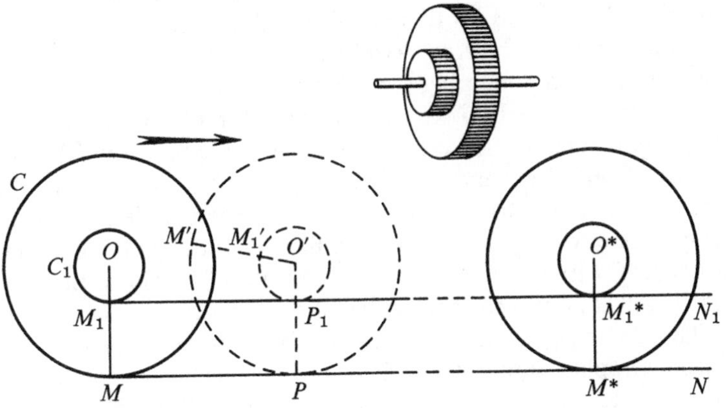

Fig. 24

circles C and C_1 at the points M and M_1, respectively, lying on the same radius OM. Since the circles are rigidly attached to each other, they will move as a unit; if one circle turns through a given angle, the other will turn through the same angle. Therefore, if the circle C rolls along the straight line MN, the circle C_1 will roll along the straight line M_1N_1. In Fig. 24 the direction in which the circles roll is indicated by the arrow, and one of the intermediate positions is depicted by dotted lines, the points M' and M_1' being the new positions of the points M and M_1. On the physical model we imagine a horizontal rail being placed under each of the cylindrical rollers; when the larger roller rolls on its rail it will cause the smaller one to roll on its own rail. Let the circle C, rolling along the straight line MN, make one complete revolution, as a result of which the point M will take up the position M^*; the circle C_1 will then also have made a complete revolution, and the point M_1 will then occupy the position M_1^* on the radius O^*M^*. O^*M^* is parallel to OM because both these radii are perpendicular to MN. We conclude that

$$MM^* = M_1M_1^*,$$

that is, that the two rolling circles go through identical distances in making a complete revolution, and that means that their circumferences are equal. As the circles C and C_1 may be taken arbitrarily, we have furnished the required proof.

Hint. We shall not decide beforehand the question as to how the reader will overcome the evident contradiction obtained; the author's deliberations on this problem will be given in Chapter 4. The following observation would seem to be useful, whatever course is taken in pondering over this sophism.

The circle is often looked upon as the limit of a succession of regular polygons inscribed in it (or circumscribed about it) as the number of their sides increases without bound.[1] This suggests that in order to obtain a clearer picture of the process undergone by the rolling circle we might instead examine a regular polygon rolling on a straight line. The greater the number of its sides, the more closely will it approach the picture of a rolling circle.

The meaning of the statement "the (convex) polygon rolls without slipping along the straight line" seems obvious; we establish a certain order of passing around the sides of the polygon, say counterclockwise, and lay one of the sides on the straight line in the starting position. We rotate the polygon about that vertex which is common to this side and the side which follows it, until the second side lies on the straight line; we then rotate about the next vertex, and so on. In short, the polygon is "tipped" from one side onto the next one, rotating each time about the vertex which is common to these sides, and as a result of this it is transported along the straight line in the direction shown.

In the case of a regular n-gon ($n = 8$ in Fig. 25), label its vertices $A_1, A_1, \ldots, A_{n-1}, A_n$ and for the initial position let the side A_1A_2 lie on the straight line along which the polygon is to roll in

[1] The term "limit" is used here deliberately, instead of the widely used "limiting position," and it has a completely precise meaning: However narrow a ring between two circles concentric with a given circle, the radius of one of the concentric circles being larger and the radius of the other smaller than the radius of the given circle (for instance, the ring may be bounded by circles of radii $R - \varepsilon$ and $R + \varepsilon$, where R is the radius of the given circle), it will be possible to find a number n such that any inscribed (circumscribed) regular polygon with more than n sides will lie completely within the ring. This must not be confused with the widely known proposition (more frequently, definition), "The length of the circumference of a circle is the limit of the sequence of lengths of the perimeters of regular inscribed (circumscribed) polygons, when . . .". As we shall see, the term "limit" has different meanings in the two cases (see below, Examples 12–14).

the direction A_1A_2. This direction is shown by the arrow in Fig. 25. In radian measure, the exterior angle of the polygon, and also the central angle, is equal to $\frac{2\pi}{n} = \frac{360°}{n}$. Therefore, it is sufficient to turn the polygon through the angle $\frac{2\pi}{n}$ about the vertex A_2 to make the side A_2A_3 lie on the straight line. After this rotation the

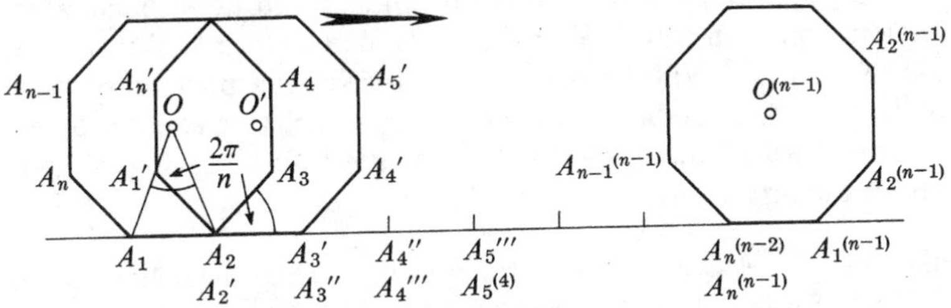

Fig. 25

center O of the polygon occupies the position O', and the vertices $A_1, A_2, A_3, \ldots, A_n$ take up the corresponding positions A_1', A_2' (coinciding with A_2), $A_3', \ldots, A_n'$. A further rotation about the vertex A_3' through the angle $\frac{2\pi}{n}$ moves the polygon $A_1'A_2'A_3' \ldots A_n'$ to the position $A_1''\,A_2''\,A_3'' \ldots A_n''$; in Fig. 25 only the vertices A_3'' (coinciding with A_3') and A_4'', both lying on the straight line, are marked. By continuing this process, we arrive after the $(n-1)$th rotation at the position $A_1^{(n-1)}A_2^{(n-1)} \ldots A_n^{(n-1)}$ with the center at the point $O^{(n-1)}$ and the side $A_n^{(n-1)}A_1^{(n-1)}$ on the straight line. As the vertex A_1 is then on the straight line once more, there is no need to continue the motion further; it is easy to see that the line segment $A_1A_1^{(n-1)}$ is equal to the perimeter of the polygon.

The reader will notice that the position of each vertex is marked by a pair of indices, the lower one showing the number of the vertex in the initial position, and the upper one, at first primes but subsequently numbers in parentheses, indicating the number of turns carried out; for instance, the symbol $A_6^{(4)}$ denotes the position of the vertex A_6 after the fourth turn. In addition to the properties inherent in the motion described in Fig. 25, it is possible to observe some accidental ones connected with the particular value chosen for n. It is suggested that the reader execute the drawing for some other, say odd, value of n, for instance, $n = 5$.

Turning once more to our sophism, we now take, instead of two concentric circles, two regular concentric n-gons whose corresponding sides are parallel, in other words, two regular polygons one of which is obtained from the other by a similarity (homothetic) transformation, with the center of similarity coinciding with the center of the second polygon. Assuming the polygons to be rigidly attached to each other, let us roll the bigger one along a straight line, as described above, and try to understand how the smaller polygon will then travel. Will the latter also be "tipped" from one side to the next? Will the perimeter of the smaller polygon be "unfolded" on the straight line as is the bigger one? (We could proceed in reverse order, rolling the smaller polygon and observing the motion of the larger one.)

Problem. We shall now formulate a problem in which the rolling circle is replaced by a rolling polygon for a different purpose. This problem shares with the subject matter of the present chapter the property of passing to the limit in some instances which require justification (see the examples which follow).

It is known that when a circle rolls along a straight line, each point on that circle moves along a curve which is called a *cycloid*. If we trace the motion of the point which is the point of contact in the initial position of the rolling circle (Fig. 26, compare with Fig. 24), its path between two consecutive

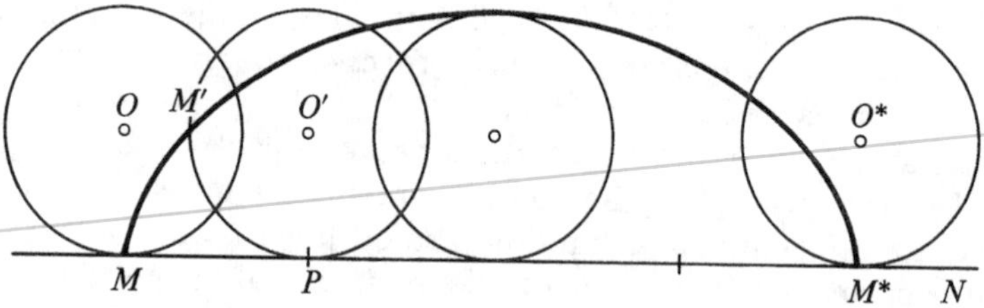

Fig. 26

points M and M^*, which, as in Fig. 24, correspond to a complete revolution of the rolling circle, will have the form of the "cycloidal arc" $MM'M^*$. By means of higher mathematics it is established that the length of this arc is equal to eight times the radius of the rolling circle, and the area included between the arc and the straight line MM^* is equal to three times the area of the circle. Our problem is to obtain these results by an elementary method. For this we follow the previous plan of replacing the rolling circle of radius R by a regular n-gon inscribed in it.

In the notation of Fig. 25, the path of the point A_1 will be composed of $(n-1)$ circular arcs, $\overset{\frown}{A_1A_1'}$ with center A_2, $\overset{\frown}{A_1'A_1''}$ with center A_3', . . . , $\overset{\frown}{A_1^{(n-2)}A_1^{(n-1)}}$ with center $A_n^{(n-2)}$; these are not shown in Fig. 25. Together these circular arcs constitute a curve which goes from A_1 to $A_1^{(n-1)}$, resembling a cycloid but differing from it by the presence of "corners" where neighboring arcs meet. As the value of n increases, the corners become smoother and the curve approaches the arc of a cycloid. It may be expected that the cycloid is the limit of this curve as $n \to \infty$. Now by means of elementary trigonometry it is easy to find the length of the path of a vertex of the polygon during one complete revolution for any value of n, since it consists of circular arcs; we can also find the area between that path and the straight line $A_1A_1^{(n-1)}$. If in the expressions obtained for the length and the area we pass to the limit as $n \to \infty$, we shall find the length to be $8R$ and the area to be $3\pi R^2$, respectively, which are the correct results.[1] However, this derivation of the formulas for the length and the area of the cycloid may be regarded as fully valid only after the passage to the limit has been justified, that is, after it has been proved that as $n \to \infty$, the values for the length and the area found for the rolling n-gon have as their limits the required length and area. It may be possible to do this within the realm of elementary mathematics, but it is certainly not easy.

The problem may be extended by also examining the paths of points inside or outside the circumference of a rolling circle and joined rigidly to it. We thereby arrive at so-called "prolate" and "curtate" cycloids. We can attempt to study these curves, too, by substituting for the rolling circle a regular inscribed polygon, and then passing to the limit.

[1] In the variant of the solution that the author has in mind, the following formulas not given in a secondary school trigonometry course are used:

$$\sin\alpha + \sin 2\alpha + \cdots + \sin k\alpha = \frac{\sin k\frac{\alpha}{2} \cdot \sin (k+1)\frac{\alpha}{2}}{\sin\frac{\alpha}{2}},$$

$$\sin^2\alpha + \sin^2 2\alpha + \cdots + \sin^2 k\alpha = \frac{2k+1}{4} - \frac{\sin(2k+1)\alpha}{4\sin\alpha},$$

$$2(\sin\alpha\sin 2\alpha + \sin 2\alpha \sin 3\alpha + \cdots + \sin(k-1)\alpha\sin k\alpha) = k\cos\alpha - \frac{\sin 2k\alpha}{2\sin\alpha}.$$

The reader can verify these identities, for instance, by induction on k. (For a study of this method see *The Method of Mathematical Induction* by I. S. Sominskii in this series.) Also involved is

$$\lim_{\omega\to 0}\frac{\sin\omega}{\omega} = 1 \qquad (\omega \text{ is the angle in radians}).$$

This formula may be found in many calculus books.

EXAMPLE 12. *The length of the hypotenuse of a triangle is equal to the sum of its two legs.*

In the right triangle ABC from the mid-point D of the hypotenuse we drop the perpendiculars DE and DF to the legs (Fig. 27, $C = 90°$); a broken line $BEDFA$ consisting of four sections is obtained, the length of which is evidently equal to the sum of the legs. We repeat this construction for each of the triangles DBE and ADF; from the mid-point of the hypotenuses DB and AD we drop perpendiculars to the legs, thereby obtaining a broken line consisting of eight sections, whose length is the same as that of the preceding one. This process may be repeated an unlimited number of times: the hypotenuse will be successively divided into 2, 4, 8, 16, . . . equal parts; a series of saw-like broken lines appears—for the sake of brevity we shall refer to them simply as "saws"—joining the points A and B and consisting, respectively, of 2, 4, 8, 16, . . . "teeth" (that is, 4, 8, 16, 32, . . . sections). All of the saws are equal in length (that is, the sums of the lengths of the sections are equal), being the sum of the lengths of the legs. As the number of sections increases, the saw will approximate more and more closely the hypotenuse AB, so that for a very large number of sections it will be difficult in practice to differentiate between the broken line with the very small sections and the continuous straight line segment (just as it is difficult to distinguish between a circle and a regular polygon with a very great number of sides inscribed in it).

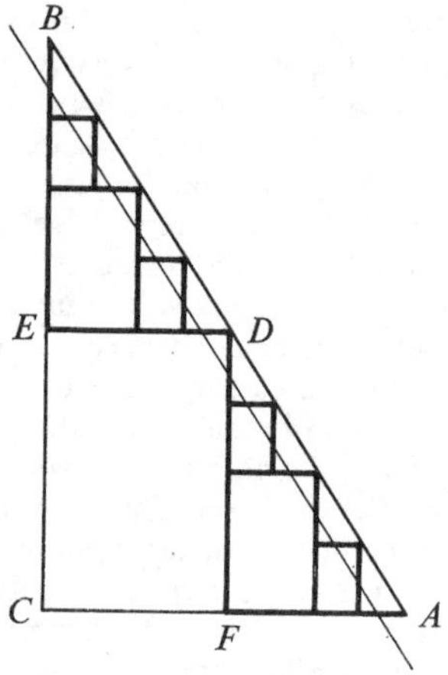

Fig. 27

We shall base a precise statement on this visual representation: The sequence of saws has the line segment AB as its limit in the sense that the greatest distance between the points of the saw and the straight line tends to zero as we consider successive members of the sequence of saws. In fact this greatest distance is simply the altitude dropped to the hypotenuse of any one of the equal right triangles which form the "teeth" of the saw, and the altitude of a tooth is less than its hypotenuse, which tends to zero. In other words, however narrow the strip between the hypotenuse AB and a straight line parallel to it which intersects the legs (Fig. 27) in the sequence of saws there exists one which, together with all those succeeding it, will fit into this strip all the way from A to B (see the footnote on page 36). But the length of all saws is the same, which means that the sequence of their lengths consists of identical numbers and its limit will be that same number, which is equal to the sum of the legs. On the other hand, the hypotenuse is the limit of the saws; hence, its length must also be the limit of the sequence of their lengths. But as a sequence cannot have two different limits, our assertion is proved.

Note 1. It is not essential that the triangle ABC be a right triangle. For an oblique triangle it is possible to construct a sequence of saws by drawing, through a point dividing one of its sides, lines parallel to the other two sides. Nor is it essential that the side be divided into 2, 4, 8, . . . equal parts; it may be divided into 2, 3, 4, 5, . . . or even into unequal parts, as long as their number increases without bound and as long as the length of the longest part tends to zero.

Note 2. The reader will perhaps look for the source of the mistake in the fact that the length of the saw remains unchanged, and that it would therefore appear to be impossible to speak of its limit. The rejoinder to this is that in mathematics we do consider sequences consisting of terms all equal to each other. According to the precise meaning of the concept "limit," this common number itself is the limit of such a sequence. However, it would not be difficult to modify our construction in such a way that the length of the saw becomes variable while everything else remains valid. It would be sufficient, for instance, to break off one of the teeth from each saw, let us say the first one counting from the point A; more strictly speaking, we replace the first two sections by a segment of the hypotenuse, beginning at A. This would decrease by one the number of sections of each saw. As before, the "damaged saws" will have the line segment AB as their limit, but their lengths will each differ from the sum of the lengths of the legs, $AC + BC$, by a very small amount and will tend to this sum as a limit.

EXAMPLE 13. *The number π is equal to* 2.

On the line segment AB as diameter construct a semicircle (Fig. 28); then bisect the segment AB and with each half as diameter

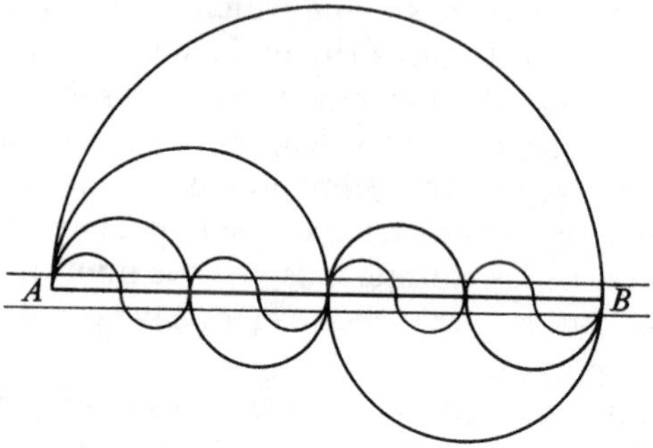

Fig. 28

construct semicircles which lie on different sides of AB. These two semicircles form a wavy line (resembling the sine curve), whose length from A to B is equal to the length of the original semicircle, that is, $\frac{\pi}{2}AB$. Each smaller semicircle is half as long as the larger one, since its diameter is half the length. Now divide the line segment AB into four equal parts and construct a wavy line consisting of four semicircles (Fig. 28), the sum of whose lengths is again $\frac{\pi}{2}AB$. We repeat this process over and over again, dividing AB into 8, 16, . . . equal parts and constructing on them semicircles which lie on alternate sides of the straight line AB. We obtain a series of wavy curves which approximate more and more closely the line segment AB. This segment represents their limit in the sense that the greatest of the distances of the points of each wavy line from the straight line AB tends to zero as we proceed to successive members of the sequence; this greatest distance is evidently equal to the radius of the semicircles of which the line is composed. (In Fig. 28 a strip between two lines parallel to AB is depicted. However narrow that strip, it is possible to find a place in our sequence from which all succeeding wavy lines will lie completely inside this strip from A to B.) But the lengths of all these wavy lines are identical and equal to $\frac{\pi}{2}AB$; this must, therefore, also be the length of the limit of these lines, that is, the length of the line segment AB. From the equality $\frac{\pi}{2} AB = AB$, we find that $\pi = 2$.

Note. This example may be supplemented by considerations analogous to notes 1 and 2 in Example 12; neither the manner in which the line segment AB is divided nor the constant lengths of the wavy lines plays any essential part. As in the preceding example, in each wavy line it would be possible to replace one of the semicircles by its diameter, and then the length of the lines would be variable. We suggest that the reader examine another variant: Instead of semicircles, that is, arcs which are subtended by right angles, we may construct arcs which are subtended by an arbitrary different angle, constant or variable according to some given law, depending on the number of divisions, but not tending to 180°. We then obtain a different value for the number π.

EXAMPLE 14. "*Schwarz's cylinder.*"

To measure the length of the arc $\overset{\frown}{AB}$ of a curved line (Fig. 29, left), we can proceed almost in the same way as when measuring

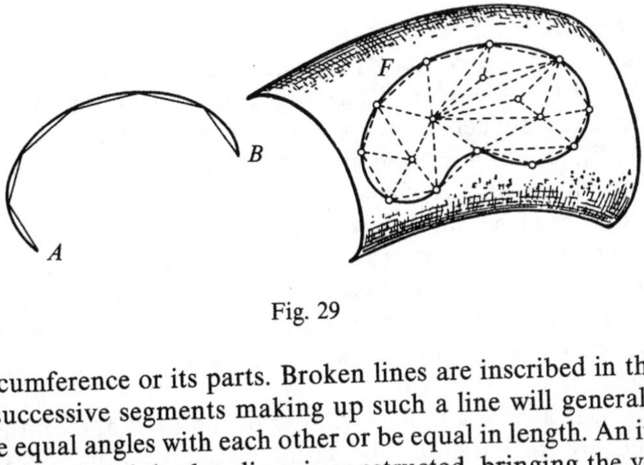

Fig. 29

a circumference or its parts. Broken lines are inscribed in the arc; the successive segments making up such a line will generally not make equal angles with each other or be equal in length. An infinite sequence of such broken lines is constructed, bringing the vertices infinitely close to each other, with the condition that as we consider successive members of the sequence, the length of the longest section of the broken line tends to zero. The limit of the sequence of lengths of these broken lines will then be the length of the original arc.

Passing on to two dimensions, we consider the analogous problem of finding the area of a figure F which lies on a curved surface (Fig. 29, right). By analogy with the arc, it appears natural to proceed as follows: In the given figure inscribe polyhedral surfaces[1] with faces of decreasing size; the area of the figure F will be the limit of the areas of these polyhedral surfaces, that is, of the sum of the areas of their faces. This is defined more precisely as follows: Inside the figure F and on its boundary take a set of points and pass planes through groups of three points to obtain a polyhedral surface with triangular faces such that no two of these faces have any common interior points and no three of them have a common edge. Construct an infinite sequence of such polyhedral surfaces

$$F_1, F_2, \ldots, F_n, \ldots$$

inscribed in the figure F, such that the length of the longest edge occurring among the sides of all the triangular faces of the surface F_n tends to zero as $n \to \infty$ and such that every point of the figure F is the limit for some sequence of points chosen successively on $F_1, F_2, \ldots, F_n, \ldots$.

It now seems to be clear that we can choose each polyhedral surface in such a way that its outside boundaries are inscribed in the boundary of the figure F (see Fig. 29). It seems almost self-evident that the sequence of areas of the polyhedral surfaces F_1, $F_2, \ldots, F_n, \ldots$ has a limit, namely, the area of the figure F. As in the preceding examples, for all practical purposes the curved surface and the polyhedral surface inscribed in it are indistinguishable when the faces of the latter become very small. At the end of the last century, however, the German mathematician G. A. Schwarz gave a simple example demonstrating that this "self-evidence" is deceptive. We shall now proceed to give an account of this example.

Consider a right circular cylinder of radius R and altitude H (Fig. 30); we shall attempt to determine the area of its lateral surface by the method set forth above. For this purpose let us divide the altitude of the cylinder into n equals parts and through these points of division pass planes perpendicular to the axis. These

[1] A polyhedral surface is inscribed in a curved surface if all of its vertices lie on the curved surface.

intersect the lateral surface of the cylinder in $n - 1$ circles, which, together with the bases, divide the lateral surface into n equal cylindrical bands. In one of these circles inscribe a regular m-gon and draw the generators of the cylinder which pass through its vertices. These generators divide each of the remaining circles into m equal parts; the points of division form the vertices of regular m-gons inscribed in these circles. The segments of the generators together with the sides of the inscribed polygons form mn identical rectangles whose vertices lie on the surface of the cylinder. One of them, $MNPQ$, is marked in Fig. 30.

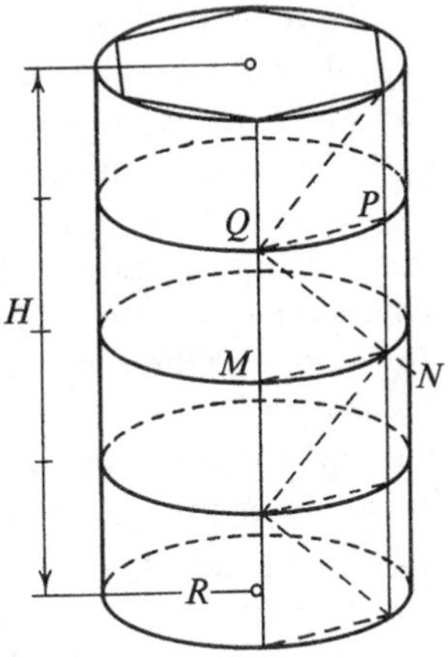

Fig. 30

Finally, dividing each rectangle into two triangles by means of a diagonal, we obtain a polyhedral surface consisting of $2mn$ identical triangular faces and inscribed in the lateral surface of the cylinder. When both m and n grow without bound, the sides of these faces tend to zero, and the distance of the points belonging to them from the lateral surface of the cylinder also tend to zero.[1]

Our inscribed polyhedral surface depends on the two indices m and n. There are infinitely many ways of choosing a sequence of our polyhedral surfaces by making one of the indices a function of the other in such a way that both indices take on integral values and simultaneously tend to infinity. For instance, we could set $m = n$, or $n = 3m$, or $m = n^2$, etc. The reader has probably already noticed that our polyhedral surfaces actually coincide with the lateral surfaces of regular m-sided prisms inscribed in the cylinder. The purpose of dividing each rectangular face of these prisms into $2n$ triangles is only to make this method of inscribing the polyhedral surfaces conform to the general scheme depicted in Fig. 29. Thus, we have before us only a somewhat complex variation of the usual derivation of the formula for the area of the lateral surface of a cylinder ($S = 2\pi RH$), that is, inscribing a regular prism and taking the limit. Up to now our arguments do not contain any sophism.

[1] The distance of any point from the lateral surface of the cylinder is just the difference between the radius of the cylinder and the distance of the point from the axis of the cylinder.

We shall now alter somewhat the method of inscribing the polyhedral surfaces. As before, we divide the altitude H into n equal parts and draw our $n - 1$ circular cross sections, which, together with the ends of the cylinder, give $n + 1$ circles. In each of these circles we inscribe a regular m-gon, but with their vertices arranged differently, namely, in such a way that the generator drawn through any vertex of the polygon inscribed in one of the circles passes halfway between vertices of the polygon inscribed in adjacent circles. For instance, in Fig. 31 the generator through P bisects the arcs $\widehat{MN}$ and $\widehat{QS}$; the figure does not show the straight lines QM and SN which are also generators. In other words, previously the polygon inscribed in any circle was obtained from the polygon inscribed in the neighboring circle simply by a parallel translation in the direction of a generator by a distance $\frac{H}{n}$, while now the translation is combined with a rotation about the center of the polygon through an angle equal to $\frac{\pi}{m}$. Arranging the regular inscribed polygons in this manner, we construct from the triangular faces a polyhedral surface (not convex!) by joining each vertex with the two vertices nearest to it on a neighboring circle. This polyhedral surface, which resembles a collapsible paper lantern in an extended position, consists of $2mn$ congruent isosceles triangles, $2m$ in each one of the n strips; it is *inscribed* in the curved surface of the cylinder in the exact sense of that word (see the footnote on page 44).

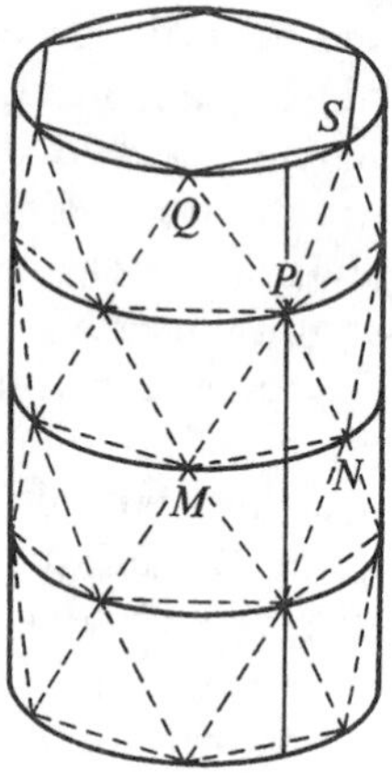

Fig. 31

To find the area of the polyhedral surface, examine one of its equal faces MNP, depicted in Fig. 31 and separately in Fig. 32 in enlarged form. Here MN is the side of the regular m-gon inscribed in the circular cross-section with center O, the points K and L are the mid-points of the arc $\widehat{MN}$ and the chord MN, respectively, and PK is a segment of the generator. Now $PM = PN$, since

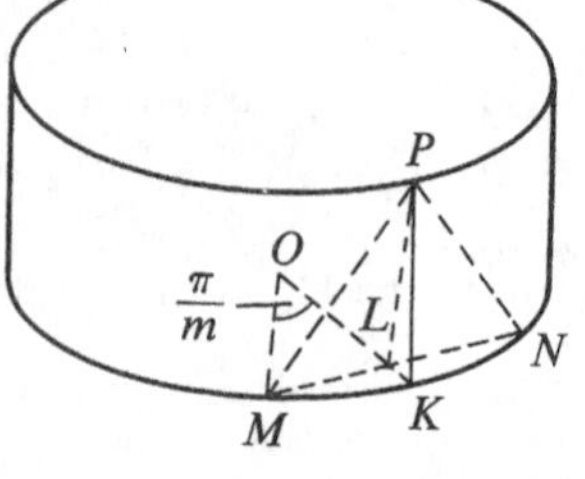

Fig. 32

these segments have equal projections onto the plane of the circle O; their projections are the equal chords KM and KN. Therefore, triangle MNP is isosceles. Its altitude PL is found from the triangle PKL, in which

$$\angle K = 90^\circ, \qquad PK = \frac{H}{n},$$

$$KL = R - OL = R - R\cos\frac{\pi}{m} = 2R\sin^2\frac{\pi}{2m}.$$

Therefore,

$$PL = \sqrt{\left(\frac{H}{n}\right)^2 + 4R^2\sin^4\frac{\pi}{2m}}.$$

And since $\frac{1}{2}MN = R\sin\frac{\pi}{m}$, we have

$$\text{area } MNP = R\sin\frac{\pi}{m}\sqrt{\frac{H^2}{n^2} + 4R^2\sin^4\frac{\pi}{2m}}.$$

Letting $S_{m,n}$ be the area of the entire polyhedral surface obtained by dividing the circle into m parts and the altitude into n parts, we find that

$$S_{m,n} = 2mn\,R\sin\frac{\pi}{m}\sqrt{\frac{H^2}{n^2} + 4R^2\sin^4\frac{\pi}{2m}}$$

$$= 2mR\sin\frac{\pi}{m}\sqrt{H^2 + 4n^2R^2\sin^4\frac{\pi}{2m}}.$$

As already remarked in another connection, it is possible in an infinite number of ways to establish a relationship between the indices m and n and obtain a sequence from the areas $S_{m,n}$. Let us examine two such ways.

1. Let $n = m^2$; that is, dividing the circle successively into 3, 4, 5, . . . parts, we divide the altitude into 9, 16, 25, . . . parts, respectively. The area of the polyhedral surface S_m (now it depends only on the index m) will be expressed by the formula

$$S_m = 2mR\sin\frac{\pi}{m}\sqrt{H^2 + 4m^4R^2\sin^4\frac{\pi}{2m}}.$$

We now take the limit as $m \to \infty$ (whereby $\frac{\pi}{m}$ and also $\sin\frac{\pi}{m}$ and $\sin\frac{\pi}{2m}$ tend to zero). Wishing to apply the last of the formulas given in the footnote on page 39, we modify the expression for S_m as follows:

$$S_m = 2\pi R \frac{\sin \frac{\pi}{m}}{\frac{\pi}{m}} \sqrt{H^2 + \frac{1}{4}\pi^4 R^2 \left(\frac{\sin \frac{\pi}{2m}}{\frac{\pi}{2m}} \right)^4},$$

after which we find that

$$\lim_{m \to \infty} S_m = 2\pi R \sqrt{H^2 + \frac{1}{4}\pi^4 R^2}.$$

This limit is evidently greater than $2\pi RH$, the generally known formula for the area of the lateral surface of a cylinder. We can obtain other values for the limit, as great as desired; for instance, putting $n = km^2$, where k is a positive integer, we have

$$\lim_{m \to \infty} S_m = 2\pi R \sqrt{H^2 + \frac{k^2}{4}\pi^4 R^2}.$$

2. Let $n = m^3$, so that as compared with the preceding method, the number of divisions along the altitude increases still more rapidly. In the formula for S_m this will merely cause the appearance of the additional factor m^2 in the second term of the sum under the radical sign. As a result, this term, and with it S_m, will tend to infinity as $m \to \infty$. This means that it is possible to establish a law for inscribing the polyhedral surfaces such that their area increases without bound and does not tend to any limit. It would appear that the area of the lateral surface of a cylinder is not a well-defined quantity.

We have arrived at an obvious absurdity and must now attempt to find the mistake.

EXAMPLE 15. *The area of a sphere of radius R is equal to* $\pi^2 R^2$.

Let us examine the hemisphere (Fig. 33) whose center lies at O, whose "equator" is q, and whose "pole" is P. This means that the radius OP is perpendicular to the plane of the equator q passing through O. Divide the equator q into a very large number of equal parts, say n, and join P to all the points of division by arcs of great circles. (Each arc is $\frac{1}{4}$ of an entire "meridian.") The hemisphere will then be divided into n very narrow spherical triangles, each of which is bounded by a small arc of the equator and two arcs of meridians. Some of these triangles are depicted in the figure; one of them, PAB, is shaded. By an increase in the number of divisions n, these spherical triangles may be made as narrow as desired, and

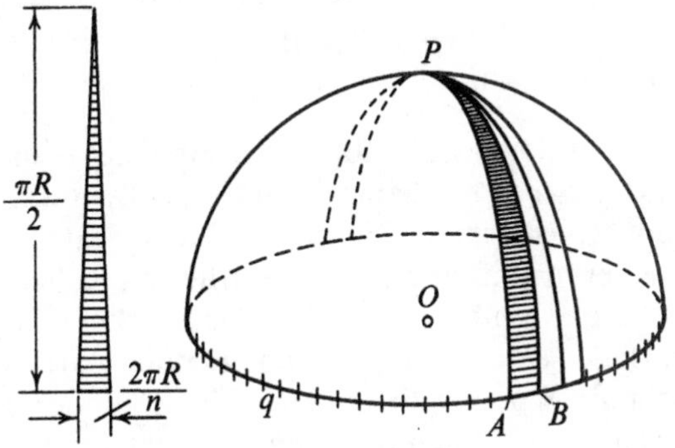

Fig. 33

the "infinitely narrow" curved triangle can be flattened out, or, as we say, "applied" to the plane with all its dimensions preserved (lengths, angles, and area). A plane triangle (isosceles) is obtained; its base is the straightened arc of length $\frac{2\pi R}{n}$, and its altitude is the straightened arc which is equal to one quarter of the circumference, that is, $\frac{\pi R}{2}$ (see the shaded triangle in Fig. 33, left). The area of such a triangle is

$$\frac{1}{2}\cdot\frac{2\pi R}{n}\cdot\frac{\pi R}{2}=\frac{1}{2n}\,\pi^2R^2;$$

consequently, the total area of all the n triangles filling up the hemisphere is equal to $\frac{1}{2}\,\pi^2R^2$, and the area of the surface of the whole sphere will be π^2R^2. This contradicts the generally-known formula according to which this area is equal to $4\pi R^2$.

4. Analysis of the Examples Given in Chapter 3

Example 11. In its classical form this sophism belongs properly not to geometry but to mechanics—more exactly to kinematics, the study of motion—inasmuch as the problem deals with a wheel moving in a certain manner. On the other hand, we can foresee that under close examination the kinematic label will prove to be purely superficial, since time plays no essential part (it is immaterial, for instance, whether the wheel rolls rapidly or slowly). The entire sophism can be stated in geometric language, as will be done later on.

Undoubtedly the weak side of our reasoning lies in the vagueness of the expression "the circle rolls without slipping along a straight line." It is only necessary to agree upon the precise meaning of this phrase in order to discover immediately that if one of the circles which are rigidly attached to each other rolls in this sense, then the other does not roll without slipping, and the sophistic proof collapses.

We shall first speak in the language of kinematics. That the circle rolls along the straight line without slipping means that the circle moves in such a way that it is in contact with the straight line at any given moment. The point of the circle at which the contact occurs thus has velocity zero at the given moment. In other words, the point of contact serves as an "instantaneous center of rotation" for the rolling circle. This means that at any given moment the velocity of any arbitrary point connected with the circle (not necessarily lying on the circumference of the circle) is that which it would possess if the circle were rotating about its point of contact. In particular, the direction of this velocity is perpendicular to the straight line which connects the given point with the point of contact. Thus, retaining the notation of Fig. 24, the direction of the velocity of a point arriving at the position M' lies along the perpendicular to the straight line $M'P$. Hence, the straight line $M'P$ is the perpendicular (sometimes called the *normal*) to the cycloid at the point M' in Fig. 26.

If, on the contrary, that point of the circle which at a given moment is in contact with the line MN has a velocity different from zero, we say that the motion takes places "with positive slippage" if this velocity is in the direction of the motion, or "with negative slippage" if it is in the opposite direction. Only if there is complete absence of any slippage is it possible to affirm that the path traveled along the straight line in a given time interval is equal to the length of the circular arc corresponding to the central angle through which an arbitrary radius of the circle has turned during this time interval. For instance, in Fig. 24 and 26, $MP = \overset{\frown}{PM'}$, and MM^* is equal in length to the whole circumference of the rolling circle. In case of positive slippage, $MP > \overset{\frown}{PM'}$; in case of negative slippage, $MP < \overset{\frown}{PM'}$.

We are now in a position to describe the different aspects of rolling in purely geometric terms, although for the sake of clarity we shall occasionally revert to the language of kinematics. Consider the line segment $MM^* = 2\pi R$ (Fig. 24 and 26), and at every point P of it construct a tangent circle with center O' lying on a given side of MM^* and having a radius R. On each circle lay off the arc $\overset{\frown}{PM'}$ equal in length to the line segment PM and on the same side of P as the line segment PM.[1] If we carry out this construction for all possible positions of the point P on the line segment MM^*, we shall say (returning to the language of kinematics, but remaining essentially in the realm of geometry) that the set of all the tangent circles is obtained as *the result of one revolution of the circle of radius R rolling without slipping along the straight line MN*, and the locus of all the points M' corresponding to different positions of the point P, will be called the *trajectory* of the point M. If in the preceding construction we replace the equality $MP = \overset{\frown}{PM'}$ by the proportionality $MP = k\overset{\frown}{PM'}$ (where k is a constant factor different from 1), we say that the circle rolls "with constant slippage coefficient k"; the slippage will be positive or negative depending on whether k is greater or less than 1.

Armed with these precise definitions, we now turn to Aristotle's wheel. From the kinematic point of view, we know that if the larger of the concentric circles depicted in Fig. 24 rolls along the

[1] This means that the arc PM' and the line segment PM lie on the same side of the diameter PO'. In the case of a rolling curve different from a circle we would say "on the same side of the normal."

straight line MN without slipping, the smaller one will not roll along the straight line M_1N_1 in the same manner. Actually, if the smaller circle were to roll without slipping, then at the moment when the common center of the circles is at the point O', the moving figure would simultaneously have two instantaneous centers of rotation P and P_1, and the velocity of the point M' would be in a direction perpendicular to both PM' and P_1M', which is impossible. Apart from this, the smaller circle rolls with positive slippage, since we always have $M_1P_1 = MP = \overset{\frown}{PM'}$, and, therefore, $M_1P_1 > \overset{\frown}{P_1M_1'}$. Conversely, if the smaller circle rolls along M_1N_1 without slipping, the larger circle carried along by it would roll with negative slippage.

We arrive at the same result if we start from a geometric definition. If the larger circle in Fig. 24 "rolls" so that in any arbitrary position $MP = \overset{\frown}{PM'}$, then $M_1P_1 = \frac{R}{r}\overset{\frown}{P_1M_1'}$, where R and r are the radii of the larger and the smaller circle, respectively. Thus, the smaller circle rolls with positive slippage of coefficient $\frac{R}{r}(> 1)$. On the other hand, if the smaller circle rolls without slipping, the larger one will roll with negative slippage of coefficient $\frac{r}{R}(< 1)$.

Hint following Example 11. Let us consider two concentric and homothetic n-gons $A_1A_2 \ldots A_n$ and $a_1a_2 \ldots a_n$ with center O. (In Fig. 34, where $n = 8$, the larger polygon retains the notation of Fig.

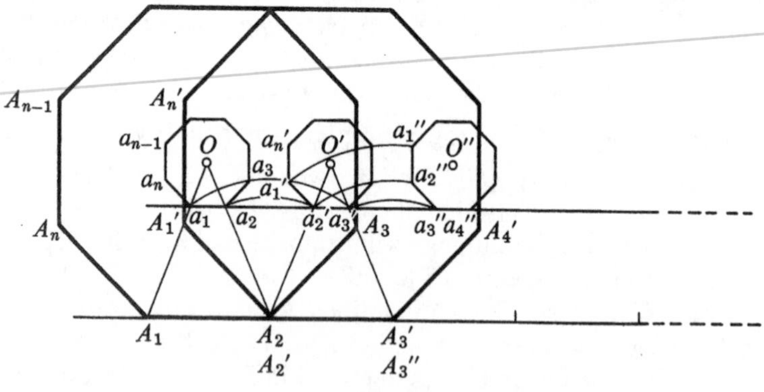

Fig. 34

25.) Let the larger polygon roll in the manner depicted in Fig. 25; at first the vertex A_2 remains stationary, serving as the center of rotation until the polygon has rotated through the angle $\frac{2\pi}{n}$. (For comparison we recall that for the rolling wheel the "lowest" point also serves as the center of rotation, but only momentarily.) As a result of such a rotation the larger polygon will lie on the straight line on another of its sides, A_2A_3, whose new position $A_2'A_3'$ in the figure forms a direct extension of the side A_1A_2; owing to this the perimeter of the polygon is "unrolled" along the straight line as the rolling proceeds.

The motion undergone meanwhile by the smaller polygon, which is attached to the larger one, will be substantially different. It will also rotate about the center A_2 through an angle of $\frac{2\pi}{n}$, as a result of which the side a_2a_3 will take up the position $a_2'a_3'$. However, the position $a_2'a_3'$ is not immediately adjacent to the position a_1a_2. In contrast to the polygon $A_1A_2\ldots A_n$, which is "tipped" from one side to the next, the polygon $a_1a_2\ldots a_n$ simultaneously "tips" and "jumps" from one position to the next. In Fig. 34 two successive positions of the larger polygon and three positions of the smaller polygon are shown; the initial sections of the paths for the vertices a_1, a_2, a_3 are depicted. Each of these sections of the path is formed from circular arcs whose magnitude in radians is $\frac{2\pi}{n}$; for instance, the path for the vertex a_2 begins with $\widehat{a_2a_2'}$ with center A_2 and $\widehat{a_2'a_2''}$ with center A_3'. As in Fig. 25, we suggest that the reader not attach any significance to some of the peculiarities of Fig. 34 which are due to the particular value $n = 8$ taken there; he should make another drawing, for instance, for $n = 5$. Owing to the "jumps," the polygon $a_1a_2\ldots a_n$, which moves along the straight line $a_1a_1^{(n-1)}$ covers a distance which is greater than its perimeter; that is an approximation of a model of "rolling with positive slippage." We suggest that the reader find out how the polygon $A_1A_2\ldots A_n$ would move if the polygon $a_1a_2\ldots a_n$ were made to roll along the straight line without slipping. It may be predicted that after each turn through the angle $\frac{2\pi}{n}$ about the vertex of the smaller polygon, the side of the larger one will partly overlap the preceding side of this polygon, with the result that the path it travels during a complete revolution will be less than its perimeter. We thus obtain an approximation of "rolling with negative slippage."

PROBLEM FOLLOWING EXAMPLE 11. It suffices to give a drawing and some intermediate results.

Fig. 35 depicts the path $A_1A_1'A_1'' \ldots A_1^{(n-1)}$ of the vertex A_1 of a rolling

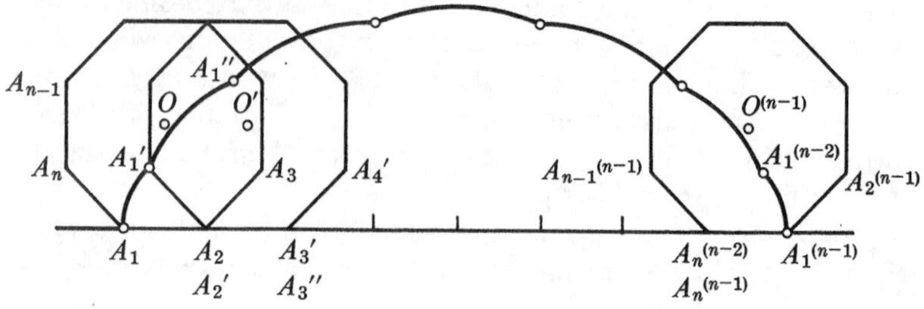

Fig. 35

polygon (in this case $n = 8$), described in greater detail on page 39. The length of this path, which consists of circular arcs, is equal to

$$\frac{2\pi}{n} \cdot 2R \left(\sin\frac{\pi}{n} + \sin\frac{2\pi}{n} + \cdots + \sin\frac{(n-1)\pi}{n}\right) = \frac{4\pi R}{n} \frac{\cos\frac{\pi}{2n}}{\sin\frac{\pi}{2n}}.$$

The area of the figure bounded by this trajectory and by the line segment $A_1A_1^{(n-1)}$ consists of (1) the areas of the circular sectors $A_2A_1A_1'$, $A_3'A_1'A_1''$, $\ldots$, $A_n^{(n-2)}A_1^{(n-2)}A_1^{(n-1)}$ with central angles of $\frac{2\pi}{n}$; the sum of these areas is equal to $\frac{4\pi R^2}{n}\left(\sin^2\frac{\pi}{n} + \sin^2\frac{2\pi}{n} + \cdots + \sin^2\frac{(n-1)\pi}{n}\right) = 2\pi R^2$; and (2) the areas of the triangles $A_1'A_2'A_3'$, $A_1''A_3'A_4''$, $\ldots$, $A_1^{(n-2)}A_{n-1}^{(n-3)}A_n^{(n-2)}$, whose sum is

$$2R^2 \sin\frac{\pi}{n}\left(\sin\frac{2\pi}{n}\sin\frac{\pi}{n} + \sin\frac{3\pi}{n}\sin\frac{2\pi}{n} + \cdots + \sin\frac{(n-1)\pi}{n}\sin\frac{(n-2)\pi}{n}\right) = nR^2 \sin\frac{\pi}{n}\cos\frac{\pi}{n}.$$

EXAMPLE 12. The logical mistake lies in the last part of the proof (preceding note 1); it is the use of the word "limit" in two entirely different senses. In one case a sequence of lines is being considered—here it is the "saws" with a variable number of "teeth"—whose points approach infinitely close to some given line. In the other case we are talking about a sequence of numbers—the lengths of the saws—which approach infinitely close to some given number. (As for the question of whether it is possible to regard the lengths of the saws as forming a sequence, see Example 12, note 2.)

The fact that a sequence of lines tends (in the first sense) to a given line gives us no basis for the conclusion that the sequence of the lengths of the lines tends (in the second sense) to the length of the given line.

We must not be confused by the fact that for very fine teeth the saw becomes practically indistinguishable from the straight line segment; this is not a geometric fact but a physical or even a physiological one, depending on the characteristics of our eyesight. (A powerful microscope would alter the situation.) But if we reinforce what we see by reason, the matter will present itself in the following form. It is true that in each small triangle ("sawtooth") the difference between the sum of the legs and the hypotenuse is insignificant, but the number of such triangles is very great and obviously even the smallest of terms may, when taken in very large numbers, give an appreciable sum. If we want to penetrate more deeply into the substance of the matter, we turn our attention to the fact that the sections of the saw approach the straight line *AB* in distance but by no means in direction; however small these sections may be, they are always alternately horizontal and vertical, whereas the hypotenuse *AB* is oblique (Fig. 27).

EXAMPLE 13. The mistake is of the same type as that in the preceding example. The sequence of wavy lines approaches infinitely close to the straight line segment, but the limit of their lengths is not the length of this segment. As before, one line (wavy in our case) approaches another in distance, but not in direction; if the line segment *AB* is horizontal, the direction of the wavy line will always oscillate between the horizontal and the vertical, however small its circular arcs.

EXAMPLE 14. Although the picture is considerably more complex than in the two preceding examples, the logical nature of the sophism is the same; the polyhedral surface does in fact approach infinitely close to the cylindrical surface, but from this it does not follow in any way that the area of the polyhedral surface approaches infinitely close to the area of the cylindrical surface. In order to obtain a clearer insight into the connection between the two approximations, we note that it is possible, even by the method of inscribing a polyhedral surface shown in Fig. 31, to obtain the correct formula for the area of the lateral surface of the cylinder. For instance, we can put $n = m$ or $n = 10m$; in general we can take the number of divisions along the altitude and along the circle as proportional to each other. For instance, for $n = 10m$ we have

$$S_m = 2mR \sin\frac{\pi}{m}\sqrt{H^2 + 400m^2R^2 \sin^4\frac{\pi}{2m}}$$

$$= 2\pi R \frac{\sin\frac{\pi}{m}}{\frac{\pi}{m}} \sqrt{H^2 + \frac{25\pi^4R^2}{m^2}\left(\frac{\sin\frac{\pi}{2m}}{\frac{\pi}{2m}}\right)^4},$$

and for $m \to \infty$ we obtain $\lim S_m = 2\pi RH = S$.

How are we to explain the fact that if we use a different law relating m and n, say, $n = m^2$, the area S_m of the polyhedral surface tends to a limit which is greater than $2\pi RH$, while for $n = m^3$ it even tends to infinity? Let us take the liberty of answering not in the language of mathematics, foregoing any pretensions of giving a proof, but singling out only what lies within the scope of visual ideas (which may, after being analyzed mathematically, become the basis of a proof). If $n = m$, or $n = 10m$, etc., the density of the division points on the circle and on the altitude will increase at the same rate. As a result, the polyhedral surface, although it is not convex, has almost vertical faces if the cylinder is placed vertically. (We suggest that the reader prove by means of Fig. 32 that the face MNP will form an angle with the horizontal plane MKN which tends to $\frac{\pi}{2}$ as $m \to \infty$.) Thus, the polyhedral surface approaches the cylindrical one not only in distance, but also in direction. A different picture is obtained when $n = m^2$, or $n = m^3$, etc.; now the division points along the altitude increase in density much faster than those on the circle. As a result, the polyhedral surface becomes significantly more "indented," and in consequence a surplus area is obtained. The triangular faces will now no longer tend to become vertical; it can be shown that for $n = m^2$, the angle PLK (Fig. 32) approaches infinitely close to some acute angle, and for $n = m^3$ it even tends to zero as $m \to \infty$; that is, the faces tend to become horizontal.

In conclusion, we shall examine one problem which arises naturally: Why is there no analogy between inscribing broken lines in a curved line and inscribing polyhedral surfaces in a curved surface? Why is it that in the first case the concentration of the vertices guarantees that the lines approximate the arcs they span not only in distance but also in direction, while in the second case there may not be an approximation in direction? Without entering into details, we note only the following facts. When one point on a curve approaches a fixed point on the curve, the limit of the straight line which connects these points will be the tangent to the curve at the fixed point. When two points on a curved surface approach a third fixed point on the surface (assume the three points are never collinear), then the plane which is determined by these points does not necessarily approach the tangent plane. To verify this it is sufficient to imagine an arbitrary circle drawn on a sphere and to take on it one fixed point and two others which approach infinitely close to the first; the plane of the three points will always be the plane of the circle.

EXAMPLE 15. Before us we have an abuse of expressions which do not have mathematical significance, such as "very large number," "very narrow triangle," "small arc," "infinitely narrow triangle"; these expressions are appropriate when attempting to give a visual description of a geometric figure—this method of description has been applied in several examples above—but they are completely unsuitable as tools for giving a proof or deriving a formula. The exact mistake lies in the assertion that an infinitely narrow triangle may be applied to a plane, that is, that it is possible to replace it by a plane triangle whose sides, angles, and areas have the same magnitude as those of the spherical triangle. In fact, it is not possible to apply any spherical triangle, however small, onto a plane in this sense. This is evident from the fact that the sum of the angles of a plane triangle is always equal to 180°, whereas for a spherical triangle it is always greater than 180°. In our example the angles A and B of the spherical triangle PAB are right angles (Fig. 33, right); if such a triangle could be applied to a plane, we would obtain a plane isosceles triangle (Fig. 33, left) with two right angles at the base.

A CATALOG OF SELECTED

DOVER BOOKS

IN ALL FIELDS OF INTEREST

A CATALOG OF SELECTED DOVER BOOKS IN ALL FIELDS OF INTEREST

STICKLEY CRAFTSMAN FURNITURE CATALOGS, Gustav Stickley and L. & J. G. Stickley. Beautiful, functional furniture in two authentic catalogs from 1910. 594 illustrations, including 277 photos, show settles, rockers, armchairs, reclining chairs, bookcases, desks, tables. 183pp. 6½ x 9¼. 0-486-23838-5

AMERICAN LOCOMOTIVES IN HISTORIC PHOTOGRAPHS: 1858 to 1949, Ron Ziel (ed.). A rare collection of 126 meticulously detailed official photographs, called "builder portraits," of American locomotives that majestically chronicle the rise of steam locomotive power in America. Introduction. Detailed captions. xi+ 129pp. 9 x 12. 0-486-27393-8

AMERICA'S LIGHTHOUSES: An Illustrated History, Francis Ross Holland, Jr. Delightfully written, profusely illustrated fact-filled survey of over 200 American lighthouses since 1716. History, anecdotes, technological advances, more. 240pp. 8 x 10¾. 0-486-25576-X

TOWARDS A NEW ARCHITECTURE, Le Corbusier. Pioneering manifesto by founder of "International School." Technical and aesthetic theories, views of industry, economics, relation of form to function, "mass-production split" and much more. Profusely illustrated. 320pp. 6⅛ x 9¼. (Available in U.S. only.) 0-486-25023-7

HOW THE OTHER HALF LIVES, Jacob Riis. Famous journalistic record, exposing poverty and degradation of New York slums around 1900, by major social reformer. 100 striking and influential photographs. 233pp. 10 x 7⅞. 0-486-22012-5

FRUIT KEY AND TWIG KEY TO TREES AND SHRUBS, William M. Harlow. One of the handiest and most widely used identification aids. Fruit key covers 120 deciduous and evergreen species; twig key 160 deciduous species. Easily used. Over 300 photographs. 126pp. 5⅜ x 8½. 0-486-20511-8

COMMON BIRD SONGS, Dr. Donald J. Borror. Songs of 60 most common U.S. birds: robins, sparrows, cardinals, bluejays, finches, more–arranged in order of increasing complexity. Up to 9 variations of songs of each species.
Cassette and manual 0-486-99911-4

ORCHIDS AS HOUSE PLANTS, Rebecca Tyson Northen. Grow cattleyas and many other kinds of orchids–in a window, in a case, or under artificial light. 63 illustrations. 148pp. 5⅜ x 8½. 0-486-23261-1

MONSTER MAZES, Dave Phillips. Masterful mazes at four levels of difficulty. Avoid deadly perils and evil creatures to find magical treasures. Solutions for all 32 exciting illustrated puzzles. 48pp. 8¼ x 11. 0-486-26005-4

MOZART'S DON GIOVANNI (DOVER OPERA LIBRETTO SERIES), Wolfgang Amadeus Mozart. Introduced and translated by Ellen H. Bleiler. Standard Italian libretto, with complete English translation. Convenient and thoroughly portable–an ideal companion for reading along with a recording or the performance itself. Introduction. List of characters. Plot summary. 121pp. 5¼ x 8½. 0-486-24944-1

FRANK LLOYD WRIGHT'S DANA HOUSE, Donald Hoffmann. Pictorial essay of residential masterpiece with over 160 interior and exterior photos, plans, elevations, sketches and studies. 128pp. 9¼ x 10¾. 0-486-29120-0

THE CLARINET AND CLARINET PLAYING, David Pino. Lively, comprehensive work features suggestions about technique, musicianship, and musical interpretation, as well as guidelines for teaching, making your own reeds, and preparing for public performance. Includes an intriguing look at clarinet history. "A godsend," *The Clarinet,* Journal of the International Clarinet Society. Appendixes. 7 illus. 320pp. 5⅜ x 8½. 0-486-40270-3

HOLLYWOOD GLAMOR PORTRAITS, John Kobal (ed.). 145 photos from 1926-49. Harlow, Gable, Bogart, Bacall; 94 stars in all. Full background on photographers, technical aspects. 160pp. 8⅜ x 11¼. 0-486-23352-9

THE RAVEN AND OTHER FAVORITE POEMS, Edgar Allan Poe. Over 40 of the author's most memorable poems: "The Bells," "Ulalume," "Israfel," "To Helen," "The Conqueror Worm," "Eldorado," "Annabel Lee," many more. Alphabetic lists of titles and first lines. 64pp. 5³⁄₁₆ x 8¼. 0-486-26685-0

PERSONAL MEMOIRS OF U. S. GRANT, Ulysses Simpson Grant. Intelligent, deeply moving firsthand account of Civil War campaigns, considered by many the finest military memoirs ever written. Includes letters, historic photographs, maps and more. 528pp. 6⅛ x 9¼. 0-486-28587-1

POE ILLUSTRATED: Art by Doré, Dulac, Rackham and Others, selected and edited by Jeff A. Menges. More than 100 compelling illustrations, in brilliant color and crisp black-and-white, include scenes from "The Raven," "The Pit and the Pendulum," "The Gold-Bug," and other stories and poems. 96pp. 8⅜ x 11. 0-486-45746-X

RUSSIAN STORIES/RUSSKIE RASSKAZY: A Dual-Language Book, edited by Gleb Struve. Twelve tales by such masters as Chekhov, Tolstoy, Dostoevsky, Pushkin, others. Excellent word-for-word English translations on facing pages, plus teaching and study aids, Russian/English vocabulary, biographical/critical introductions, more. 416pp. 5⅜ x 8½. 0-486-26244-8

PHILADELPHIA THEN AND NOW: 60 Sites Photographed in the Past and Present, Kenneth Finkel and Susan Oyama. Rare photographs of City Hall, Logan Square, Independence Hall, Betsy Ross House, other landmarks juxtaposed with contemporary views. Captures changing face of historic city. Introduction. Captions. 128pp. 8¼ x 11. 0-486-25790-8

NORTH AMERICAN INDIAN LIFE: Customs and Traditions of 23 Tribes, Elsie Clews Parsons (ed.). 27 fictionalized essays by noted anthropologists examine religion, customs, government, additional facets of life among the Winnebago, Crow, Zuni, Eskimo, other tribes. 480pp. 6⅛ x 9¼. 0-486-27377-6

TECHNICAL MANUAL AND DICTIONARY OF CLASSICAL BALLET, Gail Grant. Defines, explains, comments on steps, movements, poses and concepts. 15-page pictorial section. Basic book for student, viewer. 127pp. 5⅜ x 8½. 0-486-21843-0

THE MALE AND FEMALE FIGURE IN MOTION: 60 Classic Photographic Sequences, Eadweard Muybridge. 60 true-action photographs of men and women walking, running, climbing, bending, turning, etc., reproduced from a rare 19th-century masterpiece. vi + 121pp. 9 x 12. 0-486-24745-7

CATALOG OF DOVER BOOKS

ANIMALS: 1,419 Copyright-Free Illustrations of Mammals, Birds, Fish, Insects, etc., Jim Harter (ed.). Clear wood engravings present, in extremely lifelike poses, over 1,000 species of animals. One of the most extensive pictorial sourcebooks of its kind. Captions. Index. 284pp. 9 x 12. 0-486-23766-4

1001 QUESTIONS ANSWERED ABOUT THE SEASHORE, N. J. Berrill and Jacquelyn Berrill. Queries answered about dolphins, sea snails, sponges, starfish, fishes, shore birds, many others. Covers appearance, breeding, growth, feeding, much more. 305pp. $5\frac{1}{4}$ x $8\frac{1}{4}$. 0-486-23366-9

ATTRACTING BIRDS TO YOUR YARD, William J. Weber. Easy-to-follow guide offers advice on how to attract the greatest diversity of birds: birdhouses, feeders, water and waterers, much more. 96pp. $5\frac{3}{16}$ x $8\frac{1}{4}$. 0-486-28927-3

MEDICINAL AND OTHER USES OF NORTH AMERICAN PLANTS: A Historical Survey with Special Reference to the Eastern Indian Tribes, Charlotte Erichsen-Brown. Chronological historical citations document 500 years of usage of plants, trees, shrubs native to eastern Canada, northeastern U.S. Also complete identifying information. 343 illustrations. 544pp. $6\frac{1}{2}$ x $9\frac{1}{4}$. 0-486-25951-X

STORYBOOK MAZES, Dave Phillips. 23 stories and mazes on two-page spreads: Wizard of Oz, Treasure Island, Robin Hood, etc. Solutions. 64pp. $8\frac{1}{4}$ x 11. 0-486-23628-5

AMERICAN NEGRO SONGS: 230 Folk Songs and Spirituals, Religious and Secular, John W. Work. This authoritative study traces the African influences of songs sung and played by black Americans at work, in church, and as entertainment. The author discusses the lyric significance of such songs as "Swing Low, Sweet Chariot," "John Henry," and others and offers the words and music for 230 songs. Bibliography. Index of Song Titles. 272pp. $6\frac{1}{2}$ x $9\frac{1}{4}$. 0-486-40271-1

MOVIE-STAR PORTRAITS OF THE FORTIES, John Kobal (ed.). 163 glamor, studio photos of 106 stars of the 1940s: Rita Hayworth, Ava Gardner, Marlon Brando, Clark Gable, many more. 176pp. $8\frac{3}{8}$ x $11\frac{1}{4}$. 0-486-23546-7

YEKL and THE IMPORTED BRIDEGROOM AND OTHER STORIES OF YIDDISH NEW YORK, Abraham Cahan. Film Hester Street based on *Yekl* (1896). Novel, other stories among first about Jewish immigrants on N.Y.'s East Side. 240pp. $5\frac{3}{8}$ x $8\frac{1}{2}$. 0-486-22427-9

SELECTED POEMS, Walt Whitman. Generous sampling from *Leaves of Grass*. Twenty-four poems include "I Hear America Singing," "Song of the Open Road," "I Sing the Body Electric," "When Lilacs Last in the Dooryard Bloom'd," "O Captain! My Captain!"–all reprinted from an authoritative edition. Lists of titles and first lines. 128pp. $5\frac{3}{16}$ x $8\frac{1}{4}$. 0-486-26878-0

SONGS OF EXPERIENCE: Facsimile Reproduction with 26 Plates in Full Color, William Blake. 26 full-color plates from a rare 1826 edition. Includes "The Tyger," "London," "Holy Thursday," and other poems. Printed text of poems. 48pp. $5\frac{1}{4}$ x 7. 0-486-24636-1

THE BEST TALES OF HOFFMANN, E. T. A. Hoffmann. 10 of Hoffmann's most important stories: "Nutcracker and the King of Mice," "The Golden Flowerpot," etc. 458pp. $5\frac{3}{8}$ x $8\frac{1}{2}$. 0-486-21793-0

THE BOOK OF TEA, Kakuzo Okakura. Minor classic of the Orient: entertaining, charming explanation, interpretation of traditional Japanese culture in terms of tea ceremony. 94pp. $5\frac{3}{8}$ x $8\frac{1}{2}$. 0-486-20070-1

FRENCH STORIES/CONTES FRANÇAIS: A Dual-Language Book, Wallace Fowlie. Ten stories by French masters, Voltaire to Camus: "Micromegas" by Voltaire; "The Atheist's Mass" by Balzac; "Minuet" by de Maupassant; "The Guest" by Camus, six more. Excellent English translations on facing pages. Also French-English vocabulary list, exercises, more. 352pp. 5⅜ x 8½. 0-486-26443-2

CHICAGO AT THE TURN OF THE CENTURY IN PHOTOGRAPHS: 122 Historic Views from the Collections of the Chicago Historical Society, Larry A. Viskochil. Rare large-format prints offer detailed views of City Hall, State Street, the Loop, Hull House, Union Station, many other landmarks, circa 1904-1913. Introduction. Captions. Maps. 144pp. 9⅜ x 12¼. 0-486-24656-6

OLD BROOKLYN IN EARLY PHOTOGRAPHS, 1865–1929, William Lee Younger. Luna Park, Gravesend race track, construction of Grand Army Plaza, moving of Hotel Brighton, etc. 157 previously unpublished photographs. 165pp. 8⅞ x 11¾. 0-486-23587-4

THE MYTHS OF THE NORTH AMERICAN INDIANS, Lewis Spence. Rich anthology of the myths and legends of the Algonquins, Iroquois, Pawnees and Sioux, prefaced by an extensive historical and ethnological commentary. 36 illustrations. 480pp. 5⅜ x 8½. 0-486-25967-6

AN ENCYCLOPEDIA OF BATTLES: Accounts of Over 1,560 Battles from 1479 B.C. to the Present, David Eggenberger. Essential details of every major battle in recorded history from the first battle of Megiddo in 1479 B.C. to Grenada in 1984. List of Battle Maps. New Appendix covering the years 1967–1984. Index. 99 illustrations. 544pp. 6½ x 9¼. 0-486-24913-1

SAILING ALONE AROUND THE WORLD, Captain Joshua Slocum. First man to sail around the world, alone, in small boat. One of the great feats of seamanship told in delightful manner. 67 illustrations. 294pp. 5⅜ x 8½. 0-486-20326-3

ANARCHISM AND OTHER ESSAYS, Emma Goldman. Powerful, penetrating, prophetic essays on direct action, role of minorities, prison reform, puritan hypocrisy, violence, etc. 271pp. 5⅜ x 8½. 0-486-22484-8

MYTHS OF THE HINDUS AND BUDDHISTS, Ananda K. Coomaraswamy and Sister Nivedita. Great stories of the epics; deeds of Krishna, Shiva, taken from puranas, Vedas, folk tales; etc. 32 illustrations. 400pp. 5⅜ x 8½. 0-486-21759-0

MY BONDAGE AND MY FREEDOM, Frederick Douglass. Born a slave, Douglass became outspoken force in antislavery movement. The best of Douglass' autobiographies. Graphic description of slave life. 464pp. 5⅜ x 8½. 0-486-22457-0

FOLLOWING THE EQUATOR: A Journey Around the World, Mark Twain. Fascinating humorous account of 1897 voyage to Hawaii, Australia, India, New Zealand, etc. Ironic, bemused reports on peoples, customs, climate, flora and fauna, politics, much more. 197 illustrations. 720pp. 5⅜ x 8½. 0-486-26113-1

GREAT SPEECHES BY AMERICAN WOMEN, edited by James Daley. Here are 21 legendary speeches from the country's most inspirational female voices, including Sojourner Truth, Susan B. Anthony, Eleanor Roosevelt, Hillary Rodham Clinton, Nancy Pelosi, and many others. 192pp. 5³⁄₁₆ x 8¼. 0-486-46141-6

THE MYTHS OF GREECE AND ROME, H. A. Guerber. A classic of mythology, generously illustrated, long prized for its simple, graphic, accurate retelling of the principal myths of Greece and Rome, and for its commentary on their origins and significance. With 64 illustrations by Michelangelo, Raphael, Titian, Rubens, Canova, Bernini and others. 480pp. 5⅜ x 8½. 0-486-27584-1

HOW TO DO BEADWORK, Mary White. Fundamental book on craft from simple projects to five-bead chains and woven works. 106 illustrations. 142pp. 5⅜ x 8. 0-486-20697-1

THE 1912 AND 1915 GUSTAV STICKLEY FURNITURE CATALOGS, Gustav Stickley. With over 200 detailed illustrations and descriptions, these two catalogs are essential reading and reference materials and identification guides for Stickley furniture. Captions cite materials, dimensions and prices. 112pp. 6½ x 9¼. 0-486-26676-1

SIX GREAT DIALOGUES: Apology, Crito, Phaedo, Phaedrus, Symposium, The Republic, Plato, translated by Benjamin Jowett. Plato's Dialogues rank among Western civilization's most important and influential philosophical works. These 6 selections of his major works explore a broad range of enduringly relevant issues. Authoritative Jowett translations. 480pp. 5³⁄₁₆ x 8¼. 0-486-45465-7

DEMONOLATRY: An Account of the Historical Practice of Witchcraft, Nicolas Remy, edited with an Introduction and Notes by Montague Summers, translated by E. A. Ashwin. This extremely influential 1595 study was frequently cited at witchcraft trials. In addition to lurid details of satanic pacts and sexual perversity, it presents the particulars of numerous court cases. 240pp. 6½ x 9¼. 0-486-46137-8

VICTORIAN FASHIONS AND COSTUMES FROM HARPER'S BAZAAR, 1867–1898, Stella Blum (ed.). Day costumes, evening wear, sports clothes, shoes, hats, other accessories in over 1,000 detailed engravings. 320pp. 9⅜ x 12¼. 0-486-22990-4

THE LONG ISLAND RAIL ROAD IN EARLY PHOTOGRAPHS, Ron Ziel. Over 220 rare photos, informative text document origin (1844) and development of rail service on Long Island. Vintage views of early trains, locomotives, stations, passengers, crews, much more. Captions. 8⅞ x 11¾. 0-486-26301-0

VOYAGE OF THE LIBERDADE, Joshua Slocum. Great 19th-century mariner's thrilling, first-hand account of the wreck of his ship off South America, the 35-foot boat he built from the wreckage, and its remarkable voyage home. 128pp. 5⅜ x 8½. 0-486-40022-0

TEN BOOKS ON ARCHITECTURE, Vitruvius. The most important book ever written on architecture. Early Roman aesthetics, technology, classical orders, site selection, all other aspects. Morgan translation. 331pp. 5⅜ x 8½. 0-486-20645-9

THE HUMAN FIGURE IN MOTION, Eadweard Muybridge. More than 4,500 stopped-action photos, in action series, showing undraped men, women, children jumping, lying down, throwing, sitting, wrestling, carrying, etc. 390pp. 7⅞ x 10⅝. 0-486-20204-6 Clothbd.

TREES OF THE EASTERN AND CENTRAL UNITED STATES AND CANADA, William M. Harlow. Best one-volume guide to 140 trees. Full descriptions, woodlore, range, etc. Over 600 illustrations. Handy size. 288pp. 4½ x 6⅜. 0-486-20395-6

MY FIRST BOOK OF TCHAIKOVSKY: Favorite Pieces in Easy Piano Arrangements, edited by David Dutkanicz. These special arrangements of favorite Tchaikovsky themes are ideal for beginner pianists, child or adult. Contents include themes from "The Nutcracker," "March Slav," Symphonies Nos. 5 and 6, "Swan Lake," "Sleeping Beauty," and more. 48pp. 8¼ x 11. 0-486-46416-4

BIG BOOK OF MAZES AND LABYRINTHS, Walter Shepherd. 50 mazes and labyrinths in all–classical, solid, ripple, and more–in one great volume. Perfect inexpensive puzzler for clever youngsters. Full solutions. 112pp. 8⅛ x 11. 0-486-22951-3

PIANO TUNING, J. Cree Fischer. Clearest, best book for beginner, amateur. Simple repairs, raising dropped notes, tuning by easy method of flattened fifths. No previous skills needed. 4 illustrations. 201pp. 5⅜ x 8½. 0-486-23267-0

HINTS TO SINGERS, Lillian Nordica. Selecting the right teacher, developing confidence, overcoming stage fright, and many other important skills receive thoughtful discussion in this indispensible guide, written by a world-famous diva of four decades' experience. 96pp. 5⅜ x 8½. 0-486-40094-8

THE COMPLETE NONSENSE OF EDWARD LEAR, Edward Lear. All nonsense limericks, zany alphabets, Owl and Pussycat, songs, nonsense botany, etc., illustrated by Lear. Total of 320pp. 5⅜ x 8½. (Available in U.S. only.) 0-486-20167-8

VICTORIAN PARLOUR POETRY: An Annotated Anthology, Michael R. Turner. 117 gems by Longfellow, Tennyson, Browning, many lesser-known poets. "The Village Blacksmith," "Curfew Must Not Ring Tonight," "Only a Baby Small," dozens more, often difficult to find elsewhere. Index of poets, titles, first lines. xxiii + 325pp. 5⅝ x 8¼. 0-486-27044-0

DUBLINERS, James Joyce. Fifteen stories offer vivid, tightly focused observations of the lives of Dublin's poorer classes. At least one, "The Dead," is considered a masterpiece. Reprinted complete and unabridged from standard edition. 160pp. 5³⁄₁₆ x 8¼. 0-486-26870-5

THE LITTLE RED SCHOOLHOUSE, Eric Sloane. Harkening back to a time when the three Rs stood for reading, 'riting, and religion, Sloane's sketchbook explores the history of early American schools. Includes marvelous illustrations of one-room New England schoolhouses, desks, and benches. 48pp. 8¼ x 11. 0-486-45604-8

THE BOOK OF THE SACRED MAGIC OF ABRAMELIN THE MAGE, translated by S. MacGregor Mathers. Medieval manuscript of ceremonial magic. Basic document in Aleister Crowley, Golden Dawn groups. 268pp. 5⅜ x 8½. 0-486-23211-5

THE BATTLES THAT CHANGED HISTORY, Fletcher Pratt. Eminent historian profiles 16 crucial conflicts, ancient to modern, that changed the course of civilization. 352pp. 5⅜ x 8½. 0-486-41129-X

NEW RUSSIAN-ENGLISH AND ENGLISH-RUSSIAN DICTIONARY, M. A. O'Brien. This is a remarkably handy Russian dictionary, containing a surprising amount of information, including over 70,000 entries. 366pp. 4½ x 6⅛. 0-486-20208-9

NEW YORK IN THE FORTIES, Andreas Feininger. 162 brilliant photographs by the well-known photographer, formerly with *Life* magazine. Commuters, shoppers, Times Square at night, much else from city at its peak. Captions by John von Hartz. 181pp. 9¼ x 10¾. 0-486-23585-8

INDIAN SIGN LANGUAGE, William Tomkins. Over 525 signs developed by Sioux and other tribes. Written instructions and diagrams. Also 290 pictographs. 111pp. 6⅛ x 9¼. 0-486-22029-X

ANATOMY: A Complete Guide for Artists, Joseph Sheppard. A master of figure drawing shows artists how to render human anatomy convincingly. Over 460 illustrations. 224pp. 8⅜ x 11¼. 0-486-27279-6

MEDIEVAL CALLIGRAPHY: Its History and Technique, Marc Drogin. Spirited history, comprehensive instruction manual covers 13 styles (ca. 4th century through 15th). Excellent photographs; directions for duplicating medieval techniques with modern tools. 224pp. 8⅜ x 11¼. 0-486-26142-5

THE MALLEUS MALEFICARUM OF KRAMER AND SPRENGER, translated by Montague Summers. Full text of most important witchhunter's "bible," used by both Catholics and Protestants. 278pp. 6⅝ x 10. 0-486-22802-9

SPANISH STORIES/CUENTOS ESPAÑOLES: A Dual-Language Book, Angel Flores (ed.). Unique format offers 13 great stories in Spanish by Cervantes, Borges, others. Faithful English translations on facing pages. 352pp. 5⅜ x 8½. 0-486-25399-6

GARDEN CITY, LONG ISLAND, IN EARLY PHOTOGRAPHS, 1869–1919, Mildred H. Smith. Handsome treasury of 118 vintage pictures, accompanied by carefully researched captions, document the Garden City Hotel fire (1899), the Vanderbilt Cup Race (1908), the first airmail flight departing from the Nassau Boulevard Aerodrome (1911), and much more. 96pp. 8⅞ x 11¾. 0-486-40669-5

OLD QUEENS, N.Y., IN EARLY PHOTOGRAPHS, Vincent F. Seyfried and William Asadorian. Over 160 rare photographs of Maspeth, Jamaica, Jackson Heights, and other areas. Vintage views of DeWitt Clinton mansion, 1939 World's Fair and more. Captions. 192pp. 8⅞ x 11. 0-486-26358-4

CAPTURED BY THE INDIANS: 15 Firsthand Accounts, 1750-1870, Frederick Drimmer. Astounding true historical accounts of grisly torture, bloody conflicts, relentless pursuits, miraculous escapes and more, by people who lived to tell the tale. 384pp. 5⅜ x 8½. 0-486-24901-8

THE WORLD'S GREAT SPEECHES (Fourth Enlarged Edition), Lewis Copeland, Lawrence W. Lamm, and Stephen J. McKenna. Nearly 300 speeches provide public speakers with a wealth of updated quotes and inspiration–from Pericles' funeral oration and William Jennings Bryan's "Cross of Gold Speech" to Malcolm X's powerful words on the Black Revolution and Earl of Spenser's tribute to his sister, Diana, Princess of Wales. 944pp. 5⅜ x 8⅜. 0-486-40903-1

THE BOOK OF THE SWORD, Sir Richard F. Burton. Great Victorian scholar/adventurer's eloquent, erudite history of the "queen of weapons"–from prehistory to early Roman Empire. Evolution and development of early swords, variations (sabre, broadsword, cutlass, scimitar, etc.), much more. 336pp. 6⅛ x 9¼. 0-486-25434-8

AUTOBIOGRAPHY: The Story of My Experiments with Truth, Mohandas K. Gandhi. Boyhood, legal studies, purification, the growth of the Satyagraha (nonviolent protest) movement. Critical, inspiring work of the man responsible for the freedom of India. 480pp. 5⅜ x 8½. (Available in U.S. only.) 0-486-24593-4

CELTIC MYTHS AND LEGENDS, T. W. Rolleston. Masterful retelling of Irish and Welsh stories and tales. Cuchulain, King Arthur, Deirdre, the Grail, many more. First paperback edition. 58 full-page illustrations. 512pp. 5⅜ x 8½. 0-486-26507-2

THE PRINCIPLES OF PSYCHOLOGY, William James. Famous long course complete, unabridged. Stream of thought, time perception, memory, experimental methods; great work decades ahead of its time. 94 figures. 1,391pp. 5⅜ x 8½. 2-vol. set. Vol. I: 0-486-20381-6 Vol. II: 0-486-20382-4

THE WORLD AS WILL AND REPRESENTATION, Arthur Schopenhauer. Definitive English translation of Schopenhauer's life work, correcting more than 1,000 errors, omissions in earlier translations. Translated by E. F. J. Payne. Total of 1,269pp. 5⅜ x 8½. 2-vol. set. Vol. 1: 0-486-21761-2 Vol. 2: 0-486-21762-0

CATALOG OF DOVER BOOKS

MAGIC AND MYSTERY IN TIBET, Madame Alexandra David-Neel. Experiences among lamas, magicians, sages, sorcerers, Bonpa wizards. A true psychic discovery. 32 illustrations. 321pp. $5\frac{3}{8} \times 8\frac{1}{2}$. (Available in U.S. only.) 0-486-22682-4

THE EGYPTIAN BOOK OF THE DEAD, E. A. Wallis Budge. Complete reproduction of Ani's papyrus, finest ever found. Full hieroglyphic text, interlinear transliteration, word-for-word translation, smooth translation. 533pp. $6\frac{1}{2} \times 9\frac{1}{4}$. 0-486-21866-X

HISTORIC COSTUME IN PICTURES, Braun & Schneider. Over 1,450 costumed figures in clearly detailed engravings–from dawn of civilization to end of 19th century. Captions. Many folk costumes. 256pp. $8\frac{3}{8} \times 11\frac{3}{4}$. 0-486-23150-X

MATHEMATICS FOR THE NONMATHEMATICIAN, Morris Kline. Detailed, college-level treatment of mathematics in cultural and historical context, with numerous exercises. Recommended Reading Lists. Tables. Numerous figures. 641pp. $5\frac{3}{8} \times 8\frac{1}{2}$. 0-486-24823-2

PROBABILISTIC METHODS IN THE THEORY OF STRUCTURES, Isaac Elishakoff. Well-written introduction covers the elements of the theory of probability from two or more random variables, the reliability of such multivariable structures, the theory of random function, Monte Carlo methods of treating problems incapable of exact solution, and more. Examples. 502pp. $5\frac{3}{8} \times 8\frac{1}{2}$. 0-486-40691-1

THE RIME OF THE ANCIENT MARINER, Gustave Doré, S. T. Coleridge. Doré's finest work; 34 plates capture moods, subtleties of poem. Flawless full-size reproductions printed on facing pages with authoritative text of poem. "Beautiful. Simply beautiful."–*Publisher's Weekly*. 77pp. $9\frac{1}{4} \times 12$. 0-486-22305-1

SCULPTURE: Principles and Practice, Louis Slobodkin. Step-by-step approach to clay, plaster, metals, stone; classical and modern. 253 drawings, photos. 255pp. $8\frac{1}{8} \times 11$. 0-486-22960-2

THE INFLUENCE OF SEA POWER UPON HISTORY, 1660–1783, A. T. Mahan. Influential classic of naval history and tactics still used as text in war colleges. First paperback edition. 4 maps. 24 battle plans. 640pp. $5\frac{3}{8} \times 8\frac{1}{2}$. 0-486-25509-3

THE STORY OF THE TITANIC AS TOLD BY ITS SURVIVORS, Jack Winocour (ed.). What it was really like. Panic, despair, shocking inefficiency, and a little heroism. More thrilling than any fictional account. 26 illustrations. 320pp. $5\frac{3}{8} \times 8\frac{1}{2}$. 0-486-20610-6

ONE TWO THREE . . . INFINITY: Facts and Speculations of Science, George Gamow. Great physicist's fascinating, readable overview of contemporary science: number theory, relativity, fourth dimension, entropy, genes, atomic structure, much more. 128 illustrations. Index. 352pp. $5\frac{3}{8} \times 8\frac{1}{2}$. 0-486-25664-2

DALÍ ON MODERN ART: The Cuckolds of Antiquated Modern Art, Salvador Dalí. Influential painter skewers modern art and its practitioners. Outrageous evaluations of Picasso, Cézanne, Turner, more. 15 renderings of paintings discussed. 44 calligraphic decorations by Dalí. 96pp. $5\frac{3}{8} \times 8\frac{1}{2}$. (Available in U.S. only.) 0-486-29220-7

ANTIQUE PLAYING CARDS: A Pictorial History, Henry René D'Allemagne. Over 900 elaborate, decorative images from rare playing cards (14th–20th centuries): Bacchus, death, dancing dogs, hunting scenes, royal coats of arms, players cheating, much more. 96pp. $9\frac{1}{4} \times 12\frac{1}{4}$. 0-486-29265-7

MAKING FURNITURE MASTERPIECES: 30 Projects with Measured Drawings, Franklin H. Gottshall. Step-by-step instructions, illustrations for constructing handsome, useful pieces, among them a Sheraton desk, Chippendale chair, Spanish desk, Queen Anne table and a William and Mary dressing mirror. 224pp. 8⅛ x 11¼. 0-486-29338-6

NORTH AMERICAN INDIAN DESIGNS FOR ARTISTS AND CRAFTSPEOPLE, Eva Wilson. Over 360 authentic copyright-free designs adapted from Navajo blankets, Hopi pottery, Sioux buffalo hides, more. Geometrics, symbolic figures, plant and animal motifs, etc. 128pp. 8⅜ x 11. (Not for sale in the United Kingdom.) 0-486-25341-4

THE FOSSIL BOOK: A Record of Prehistoric Life, Patricia V. Rich et al. Profusely illustrated definitive guide covers everything from single-celled organisms and dinosaurs to birds and mammals and the interplay between climate and man. Over 1,500 illustrations. 760pp. 7½ x 10⅛. 0-486-29371-8

VICTORIAN ARCHITECTURAL DETAILS: Designs for Over 700 Stairs, Mantels, Doors, Windows, Cornices, Porches, and Other Decorative Elements, A. J. Bicknell & Company. Everything from dormer windows and piazzas to balconies and gable ornaments. Also includes elevations and floor plans for handsome, private residences and commercial structures. 80pp. 9⅜ x 12¼. 0-486-44015-X

WESTERN ISLAMIC ARCHITECTURE: A Concise Introduction, John D. Hoag. Profusely illustrated critical appraisal compares and contrasts Islamic mosques and palaces–from Spain and Egypt to other areas in the Middle East. 139 illustrations. 128pp. 6 x 9. 0-486-43760-4

CHINESE ARCHITECTURE: A Pictorial History, Liang Ssu-ch'eng. More than 240 rare photographs and drawings depict temples, pagodas, tombs, bridges, and imperial palaces comprising much of China's architectural heritage. 152 halftones, 94 diagrams. 232pp. 10¾ x 9⅞. 0-486-43999-2

THE RENAISSANCE: Studies in Art and Poetry, Walter Pater. One of the most talked-about books of the 19th century, *The Renaissance* combines scholarship and philosophy in an innovative work of cultural criticism that examines the achievements of Botticelli, Leonardo, Michelangelo, and other artists. "The holy writ of beauty."–Oscar Wilde. 160pp. 5⅜ x 8½. 0-486-44025-7

A TREATISE ON PAINTING, Leonardo da Vinci. The great Renaissance artist's practical advice on drawing and painting techniques covers anatomy, perspective, composition, light and shadow, and color. A classic of art instruction, it features 48 drawings by Nicholas Poussin and Leon Battista Alberti. 192pp. 5⅜ x 8½. 0-486-44155-5

THE ESSENTIAL JEFFERSON, Thomas Jefferson, edited by John Dewey. This extraordinary primer offers a superb survey of Jeffersonian thought. It features writings on political and economic philosophy, morals and religion, intellectual freedom and progress, education, secession, slavery, and more. 176pp. 5⅜ x 8½. 0-486-46599-3

WASHINGTON IRVING'S RIP VAN WINKLE, Illustrated by Arthur Rackham. Lovely prints that established artist as a leading illustrator of the time and forever etched into the popular imagination a classic of Catskill lore. 51 full-color plates. 80pp. 8⅜ x 11. 0-486-44242-X

HENSCHE ON PAINTING, John W. Robichaux. Basic painting philosophy and methodology of a great teacher, as expounded in his famous classes and workshops on Cape Cod. 7 illustrations in color on covers. 80pp. 5⅜ x 8½. 0-486-43728-0